Global Environmental Governance

Karl Bruckmeier

Global Environmental Governance

Social-Ecological Perspectives

Karl Bruckmeier
Higher School of Economics
National Research University
Moscow, Russia

ISBN 978-3-030-07453-1   ISBN 978-3-319-98110-9   (eBook)
https://doi.org/10.1007/978-3-319-98110-9

Cover illustration: valio84sl/gettyimages

This Palgrave Macmillan imprint is published by the registered company Springer Nature Switzerland AG
The registered company address is: Gewerbestrasse 11, 6330 Cham, Switzerland

Preface

The review of global environmental governance in this book aims to identify new ways of knowledge use in the transition to sustainability. The book is the third and final part of the project of reconstructing social ecology as a new interdisciplinary science of nature–society relations in historical, theoretical and practical perspectives. Problems of knowledge synthesis, transfer and application are discussed with the aim to support a renewing of the sustainability process. Beyond visions of sustainable resource management and empirical knowledge from research in ecology and policy analysis, science, the governance processes depend on knowledge gained from interdisciplinary and theoretical systems analyses of social–ecological systems.

The social–ecological transformation of modern society to a sustainable future society is a long-term process of several generations. When the global discourse of sustainable development began, about thirty years ago, the nature of the changes on the way to sustainability was not clear. Social, political, economic and ecological changes that affect each other need to be analysed. The formal institutions and international regimes for environmental policy, most important the climate and the sustainability regimes, depend from further social and ecological processes and from international power relations that block the transformation to sustainability at regional, national and global levels. Knowledge generated in interdisciplinary social–ecological research about modern society and its interaction with nature, about the modification of nature through humans in the epoch called "the anthropocene", is to large degrees unused in the governance processes. The biophysical interaction and feedback in material cycles, the processes of global environmental change, the socially structured interaction between nature and society at large require more and more intensive forms of knowledge governance: global environmental governance and social–ecological transformation become conflict and knowledge intensive processes.

Moscow, Russia   Karl Bruckmeier
July 2018

Acknowledgements

For the work with this book I benefited from discussions in several human ecological conferences and in my teaching and research at the National Research University—Higher School of Economics at Moscow. The persons to be named more concretely include the ones who worked with the manuscript, reading, reviewing or editing it. Earlier versions of several chapters were read critically by Iva Miranda Pires and Ana Velasco Arranz. I am very grateful for their comments that helped to clarify the writing. I want to thank the editorial team and the anonymous reviewers at Palgrave Macmillan for their editing work and the comments that helped to structure the chapters more clearly. For the book project no financial support was provided. The personal communication with colleagues and students, in teaching and research, helped to rethink the processes of global environmental governance and of sustainable development as conflict-based processes, as knowledge processes, and as processes of social-ecological transformation. These are the guiding ideas of this book that developed in many years of research on the use and management of natural resources.

Contents

Abbreviations

BRICS	Brazil, Russia, India, China, South Africa
HANPP	Human Appropriation of Net Primary Production
IAASTD	International Assessment of Agricultural Science and Technology for Development
IPCC	Intergovernmental Panel on Climate Change
OECD	Organization for Economic Co-operation and Development
UN	United Nations
UNCED	United Nations Conference on Environment and Development
UNCHE	United Nations Conference on the Human Environment
UNEP	United Nations Environment Programme

List of Tables

List of Boxes

1

Introduction

This book analyses the significance of social–ecological knowledge for the purposes of critical review and renewal of global environmental governance and the sustainability process. The book is the third and final part in a series on the development of social ecology as a new interdisciplinary science of nature–society relations The first two books, also from this publisher, were the preparatory study of "Natural Resource Use and Global Change" from 2013, describing the development of a new interdisciplinary social ecology, and the theoretical reconstruction of an emerging social–ecological theory of nature–society interaction in "Social–Ecological Transformation: Reconnecting Society and Nature" from 2016. The new publication builds on the earlier ones, taking up further themes and problems to be analysed in social ecology, thus completing the project of reconstructing social ecology in historical, theoretical, and practical perspectives. In difference to the first two books, this one focusses on the practical problems of knowledge synthesis, transfer and application for the solution of global environmental and resource use problems in global environmental governance. It does so by showing, how the practical governance processes depend from scientific input, not only of empirical knowledge, furthermore

© The Author(s) 2019
K. Bruckmeier, *Global Environmental Governance,*
https://doi.org/10.1007/978-3-319-98110-9_1

of interdisciplinary and theoretical knowledge from system analyses of social–ecological systems.

The perspective of analysis in the following chapters is interdisciplinary, showing, that the discussion of governance cannot be limited to empirically investigated problems and deficits of negotiation and implementation of environmental policies, to questions of compliance and coercion. Environmental policy and governance are analysed here in the context of knowledge about global social and environmental change as it is created in social ecology. The overarching aim of the analysis is to identify possible pathways of social–ecological transformation of modern society to a more environmentally sustainable future society that encompasses the national societies in a gradually emerging global society. The transformation processes include political and further processes that can only to a limited degree be managed, controlled and regulated: processes of *global social change*, especially economic globalisation and the emergence of global societal and economic systems; and *global environmental change*, especially climate change, biodiversity reduction, and land use change, including urbanisation. Only when these broader processes are analysed and understood, so the guiding assumption of the following analyses, can global policy- and governance processes be developed and deal successfully with the regulation of such complex processes of change that result from the interaction of modern society with nature.

Recent theoretical approaches for analysing global change include the emerging social–ecological theory of nature–society interaction, the emerging theory and approach of earth system governance, and similar interdisciplinary approaches in environmental research, especially sustainability science, political ecology and world system analysis. All of these approaches create knowledge for the reformulation and renewal of sustainable development that became the guiding idea in national and international environmental policies in the last three decades, without resulting so far in significant progress towards sustainability. The progress achieved is limited to national and regional levels of policy and natural resource management, but hardly advancing at international and global levels, as the Millennium Ecosystem Assessment (2005) and later global assessments have shown. The global transformation to sustainability is, after reviewing the interdisciplinary knowledge about global change, reconceptualised. New approaches of global environmental governance that can support the transformation to sustainability better than the limited knowledge from empirical policy analyses, require inter- and transdisciplinary knowledge integration as it is discussed in social ecology (Hummel et al. 2017; Görg et al. 2017). According to these and further critical discussions of environmental governance (Peterson 2018), it is especially social-scientific knowledge and non-scientific, practical and local knowledge that is insufficiently used in governance processes which remain, also in global governance, multi-scale approaches of integrated local, national and global governance. The knowledge used needs to be continually reviewed, synthesised, and translated for policies which include other knowledge use practices than environmental policies so far: knowledge integration and synthesis, cooperation of governmental and non-governmental actors, policies as experiments, collective learning, multi-scale and adaptive governance. Whereas so far power relations were a dominant theme in discussing and reviewing environmental governance, in future more attention needs to be given to knowledge processes and practices.

Environmental politics and governance deal with problems which never before have been subject of political action or regulation; these problems are in their complexity beyond the limits of specialised knowledge, or even beyond the limits of presently available scientific knowledge at all. What needs to be learned for environmental governance needs to be learned on the way of transformation, in new logics and with new methods of scientific research that have for the first time be described in social ecology and approaches like sustainability science. Environmental policy and governance become paradigmatic examples of dealing with limits of knowledge and ignorance. Governance is seen in social ecology from another, interdisciplinary perspective, and with other aims, that of supporting the social–ecological transformation of modern society to a more environmentally sustainable future society. In this interdisciplinary perspective, global environmental governance appears as a field of deficits where new forms of knowledge use and knowledge practices are required to deal with deficits of scientific and practical knowledge (reflected in the debates of "post-normal science" or sustainability science), deficits of methodological and theoretical reflection, of discussion of theory–practice relations, and of science–policy relations. For its renewal and improvement, global environmental governance needs to apply theoretical and interdisciplinary knowledge that is usually not used in the policy-centred debates; the difficulties are that of how to deal this new knowledge in the policy and governance discourses and practices, that are neither prepared for nor open for such broadening of the knowledge use.

Global environmental governance is understood here in a broader sense than in the political debates of global governance (Meadowcroft 2007; Moomaw et al. 2017). The broadening includes two new elements of a critical analysis of complex systems: (a) an analysis of global political and economic power relations and structures institutionalised in the modern economic world system that influences in manifold ways the policy processes; and (b) an analysis of the complex interactions between social and ecological systems at various scale levels, from social to global. The latter interactions are the key for understanding the relations between environmental governance and the broader systemic processes of global social and environmental change and transformation towards sustainability (Krausmann et al. 2009).

With the first step of broadening the perspective, the analysis goes beyond analyses of the formal institutions and international regimes for environmental policy that make the bulk of global environmental policies, with climate and sustainability regimes as overarching processes. The aim is to identify the multiple forms of dependence of global politics and governance from social and ecological processes, from international power relations, and from the knowledge practices in political processes and institutions that support, select and limit simultaneously the success of transformation to sustainability at regional, national and global levels. The selective knowledge use, the distortions, the controversies, the limits, the supporting and the blocking factors in global agency should be understood better (International Social Science Council 2010). The first step connects with the second step of critical analyses, the use of interdisciplinary social–ecological knowledge to understand the non-social, the ecological processes that influence, determine or limit the processes of global environmental governance.

This knowledge is generated and structured in theoretical and empirical interdisciplinary research in the environmental sciences, as knowledge about modern society and its interaction with nature, about the modification of nature through humans in the epoch now called "the anthropocene". The knowledge includes that about complex global ecological processes, biophysical interaction and feedback in material cycles, processes of global environmental change, the socially structured interaction between nature and society at large. In this second step of knowledge generation and integration, environmental governance is reconnected with the social–ecological transformation to sustainability as the overarching process.

As shorthand formulation for this broad perspective the inexact term of *"integrated global governance"* is used; it describes interdisciplinary social–ecological forms of analysis in which multi-scale processes, inter-systemic relations, interdependencies and entanglements between social (political, economic) systems and ecological systems are included. This perspective has some similarities with the term of earth systems governance used in the global governance project (Biermann 2007). The differences to this perspective, important for the reasoning in this book, include a more critical system analysis of interacting social and ecological systems: less policy-centred than earth system governance and more critical with regard to the underlying normative ideas and the idealism that influence earth governance thinking and large parts of environmental thinking of environmental movements. To make visible, how governance is dependent on complex social and ecological systems and their transformation is a precondition to develop and renew environmental governance that has lost dynamics after decades of struggling with insufficiently understood or ignored processes at the interfaces of nature and society that materialise in global environmental change. The resulting knowledge problems discussed in the following chapters can be summarised as follows.

1. *Inter- and transdisciplinary disciplinary knowledge integration connecting environmental research and political practice* are required for the further development of global environmental governance, where the knowledge needs to be applied for implementing global action programmes in climate policy, biodiversity management, land use

policy, water management. In research, in science communication, and in the application of available knowledge in global policy and governance the practical difficulty met is that of manifold, competing and contradicting forms of knowledge in environmental research and contradicting interpretations of global change—more than a lack of knowledge that is often diagnosed and highlighted as difficulty of knowledge integration and application, for example, in the hypothesis of "post-normal science" (Funtowicz and Ravetz 1993). That environmental regulation shows, so far, little success, is discussed as a consequence of the complexity of interacting global systems and processes, as policy failure resulting from power asymmetries between and within countries and from the bureaucratic organisation of policy processes, as deficits of coordination of policies resulting from vested interests of powerful actors, and as contradictions between the processes of sustainable development and economic globalisation or deregulation of markets (Martens and Raza 2010). To discuss the environmental problems in terms of management and governance, cooperation and coordination of policies, as action and implementation deficits, implies, however, simplifications, selective knowledge use, reductions of the complexity of processes in interacting social and ecological systems, and as consequence misunderstandings of environmental problems as such that can be solved in policy processes, through policy reforms, clean-up, repair of damages and restoration of ecosystems. Such attempts of problem-solving through reforms and legislation have inherent limits and contradictions that become visible through a broadening of the knowledge perspectives and the analyses of systems from which policies depend.

2. *Interdisciplinary broadening of the knowledge basis for environmental governance* is the main argument developed in this book, to be achieved through synthesis of knowledge for environmental agency and for social–ecological transformation, accounting for the complex processes in modern society and economy, in the capitalist world system and its world ecology that remain outside the agendas of policy reforms. This broadening implies to feed in the governance processes knowledge from social-scientific research and from theoretical analyses of societal and ecological systems and their interaction. Briefly said: The renewal of environmental governance requires that, what in the debates of the

International Social Science Council (2010) has been called "transformative cornerstones of social science research for global change", highlighting the key concept of transformation. Transformation is a theoretical term for specific forms of social and societal change that developed from institutional economics (Polanyi 1944) where it described the large, systemic change of the economy in England with the introduction of a modern market economy in the building of modern capitalism, implying also changes of social class and property relations that formed the core of the modern economic world system. The term has since then broadened its meaning and developed with recent social–ecological research to the concept of social–ecological transformation, also described as "another great transformation" or, in more theoretical terms, as transformation of socio-metabolic regimes (Krausmann et al. 2009). It is the core term of the following analyses, which differs in its meaning from the diffuse terms of social change, development, or transition; these imply heterogeneous forms of change, but rarely change in the meaning of transformation of societal and economic systems. This transformation is not limited to a political process, although it implies political regulation and change; it is a broader systemic process which refers to connected changes of coupled social and ecological systems as they are discussed with the term of sustainable development, or in more theoretical form, as "co-evolution" (Norgaard) of social and ecological systems.

3. *The social and ecological processes of global change, the economic globalisation process, and the social changes in modern society* since the collapse of East European socialism resulted in a manifold social, economic, political and environmental problems and crises, including the interconnected global problems of poverty, malnutrition and food security, access to water and energy problems, global warming and climate change, biodiversity reduction and land use change through deforestation, desertification and urbanisation, old and new social forms of inequality, political conflicts and violence in natural resource use, escalating to new forms of civil wars and wars as "climate wars" (Welzer, Dyer). Consequences of the multiple and interacting problems include the rapidly deteriorating state of ecosystems, transgressing of planetary boundaries of resource use, failing of environmental policies, the difficulties of sustainable development and

global environmental governance. Conflict mitigation at various levels, of "ecological distribution conflicts" (Martinez-Alier) becomes a main component of global environmental agency and governance. A renewal of governance through new interdisciplinary knowledge production and syntheses in environmental research, in search for ways out of the combined problems and crises, implies the shifts of perspectives and temporal horizons of governance discussed in the following chapters. Short-term political and managerial perspectives, the near future for which planning is possible, prevail in environmental policies: in attempts to improve natural resource use and management through participatory approaches, civil society action, adaptive management. These are only minor parts of the newly conceptualised global environmental governance. What is not or insufficiently accounted for in the short-term policy perspectives is the complexity of governance, the linkages between the problems, crises, and processes of global social and environmental change. This is attempted in the following reviews and knowledge syntheses.

4. With the *integration of knowledge from social and natural scientific research emerges a picture of the complexity of social and environmental change* to which politics, governance and regulation do not react sufficiently in the perspectives and the pragmatism of policy processes, policy evaluation, social and environmental impact assessment, joint learning and participation of stakeholders. These processes are necessary and become meaningful only as initiations of further changes described here as social–ecological transformation. The broadening of policy and governance processes, making them more complex, evokes quickly questions of the limits of governance; however, it is necessary to deal with the connections between global environmental and economic problems, problems in knowledge production and application, in collective action and governance. The renewed governance with a long time-perspective of several generations has to deal with the consequences of economic globalisation and global environmental change; with the social degradation, marginalisation and impoverishment of large parts of the population in all countries; with the relations between national or governmental and international or global politics; with the self-destruction of the growth-society that

leaves the essential processes of resource use, regulation and governance to market-inherent processes of technological innovations and market-oriented policies. It is critically discussed that governments and their policies locked themselves with the deregulation of market processes in the prison of the markets that cannot more than reacting to short-term requirements of production, exchange and consumption. But the governance or "navigation" of long-term social–ecological transformation processes is so far an unknown cybernetic knowledge practice. The conclusions from the diagnosed complexity of problems and processes remain controversial: How to broaden the perspectives, build and modify governance and management processes, and initiate another great transformation of economy and society? Yet, it cannot be avoided to think about a necessary transformation of knowledge practices that can be described in simple form through changes from short-term to long-term perspectives, rationalities, and governance perspectives. "In the long run we are all dead" (Keynes) has served so far as a symbol of the short-termism in economic thinking where the long-term future is less unknown than horrifying. "Nature's economy", as one of the meanings of the term ecology, is now on the global agenda for nearly half a century, without that it was possible, with all scientific knowledge, to develop and practice such long-term perspectives of collective action and governance. The reasons for this incapacity need to be analysed further, with the reflection and discussion of possibilities to design governance approaches in terms of the distant future, allowing to go beyond the ecological modernisation-thinking with the establishment of a second, ecological, rationality in the economic processes; the self-destructive economic rationality of growth as the systemic imperative of modern world system will not be corrected through the coexistence of an economic and an ecological rationality.

5. *Interdisciplinary knowledge production and transdisciplinary integration of scientific and practical knowledge are efforts to deal with the multifaceted crisis of modern society*, in methodological debates about new, transdisciplinary knowledge production, about a science of complexity and analysis of complex systems, transformative science and literacy, conviviality, true sustainability and efforts to restart the discourse

of a critical theory of capitalist society and world ecology (discussed in the following chapters). Not all of the new forms of knowledge practices in science and politics are convincing; the postmodernist discourse came to a dead end with its cultural relativizing and fragmenting of scientific knowledge production; "Big Data" became not only a description for voluminous datasets that require the new forms of software for data processing, but a new, internet-based ideology of knowledge production, that will not be able to make theories as forms of knowledge integration superfluous. What is necessary to deal with the problems of knowledge synthesis for long-term action perspectives, is: to adapt the epistemological, theoretical and methodological thinking in science that is programmed for disciplinary practices of knowledge production, to the requirements of interdisciplinary thinking, research and knowledge synthesis. An attempt to go further in this direction is made in the second part of this book, where epistemological and methodological questions of knowledge bridging and synthesis are discussed.

The search for new forms of knowledge generation and application for environmental governance is done in all chapters of this book. Interdisciplinary and critical perspectives as they develop in the discourses of social, political and human ecology imply a self-critique of scientific knowledge production; they are not meant to replace the old knowledge practices, but to combine, modify and broaden them. With the interdisciplinary environmental discourse and with the interdisciplinary research fields develop critiques of the "false totality" of disciplinary theories of modern society that do not account for the relations between society and nature. In interdisciplinary analyses and theories, society is seen in larger contexts, in its relations to nature, with regard to demographic processes, (over)use of natural resources, environmental pollution, and global change in the earths' ecosystems. Global environmental governance is the theme of this interdisciplinary investigation that shows at the end another picture of global governance as found in the growing specialised policy research: a view of global environmental governance as way beyond policy and governance with the help of governance, requiring new theoretical framing of knowledge that helps to create global agency and new knowledge practices, avoiding the selectivity and naivety that motivates environmental movements and governance, in failing efforts of building consensus, participation, cooperation and global solidarity. Today, when the scarcest resource is the social resource of agency, of collective action capacity to transform society and economy, when the maladaptive forms of social and environmental change drive the global economy and society towards self-destruction, it seems necessary to break the roadblocks to "our common future" with new forms of knowledge generation, synthesis and application.

References

Biermann, F. (2007). 'Earth System Governance' as a Crosscutting Theme of Global Change Research. *Global Environmental Change, 17*(326), 337.

Funtowicz, S., & Ravetz, J. (1993). Science for the Post-normal Age. *Futures, 25*(7), 739–755.

Görg, C., Brand, U., Haberl, H., Hummel, D., Jahn, T., & Liehr, S. (2017). Challenges for Social-Ecological Transformations: Contributions from Social and Political Ecology. *Sustainability, 9*, 1045. https://doi.org/10.3390/su9071045.

Hummel, D., Jahn, T., Keil, F., Liehr, S., & Stiess, I. (2017). Social Ecology as Critical, Transdisciplinary Science—Conceptualizing, Analyzing and Shaping Societal Relations to Nature. *Sustainability, 9*, 1050. https://doi.org/10.3390/su9071050.

International Social Science Council (ISSC). (2010). *Transformative Cornerstones of Social Science Research for Global Change*. Paris. www.world-socialscience.org.

Krausmann, F., Fischer-Kowalski, M., Schandl, H., & Eisenmenger, N. (2009). The Global Socio-Metabolic Transition: Past and Present Socio-Metabolic Profiles and Their Future Trajectories. *Journal of Industrial Ecology, 12*(5–6), 537–656.

Martens, P., & Raza, M. (2010). Is Globalisation Sustainable? *Sustainability, 2*, 290ff.

Meadowcroft, J. (2007). Who Is in Charge Here? Governance for Sustainable Development in a Complex World. *Journal of Environmental Policy & Planning, 9*(3–4), 299–314.

Millennium Ecosystem Assessment. (2005). *Ecosystems and Human Well-Being: Findings of the Scenarios Working Group, Millennium Ecosystem Assessment.* Washington, DC, USA: Island Press.

Moomaw, W. R., Bandary, R. R., Kuhl, L., & Verkoijen, P. (2017). Sustainable Development Diplomacy: Diagnostics for the Negotiation and Implementation of Sustainable Development. *Global Policy, 8*(1), 73–81.

Peterson, M. J. (2018). *Contesting Knowledge in International Environmental Governance.* London and New York: Routledge.

Polanyi, K. (1944). *The Great Transformation.* New York: Farrar & Rinehart.

Part I

The Global Environmental Situation

2

Environmental Change: Human Modification of Nature—Social and Environmental Consequences

Anthropogenic modification of nature began in early human societies of hunters and gatherers at low levels; it intensified and accelerated during the last 10,000 years, with the development of agriculture and industry, reaching the levels of global environmental change only during the short epoch of the industrial society. The historically changing forms of nature–society interaction are described below for different societies and cultures in human history, in two perspectives: a (Neo-) Malthusian perspective, where population growth is seen as a main factor for modification and destruction of nature, and a social–ecological perspective, with more complex analyses of societal change and transition. Human–nature interactions are described in an integrated conceptual framework in three complementary perspectives: human relations with nature where humans in the sense of the biologically are perceived as species and as individuals (a "Malthusian" perspective of population growth, biological reproduction, subsistence, and limits of natural resource use); social relations with nature for social groups and classes, economic systems, modes of production and societies as total social units (guided by the theoretical concept of societal metabolism); and culturally and cognitively varying social—scientific, cultural, religious

© The Author(s) 2019
K. Bruckmeier, *Global Environmental Governance*,
https://doi.org/10.1007/978-3-319-98110-9_2

or spiritual—constructions of society–nature relations (symbolic forms and reflections about humans and nature).

2.1   Complex Social–Ecological Systems and Their Change

Nature, ecosystems and coupled social and ecological systems include heterogeneous forms of complexity in their development, change and interaction. The salient quality of complex social and ecological systems, the creation of emergent properties, appears at the level of the whole system, although it is created through the interaction of the parts of the system. For nature and society, interacting through coupled social and ecological systems, the systemic mechanisms for maintaining structures and processes, cannot be described and explained with one and the same terminology and methodology as, for example, assumed in some versions of systems theory, levelling the specific differences between nature and society. Also when nature and society are transformed through humans and create emergent properties such as anthropogenic climate change, they still function as heterogeneous systems with different capacities, cannot replace each other. A consequence of complex systems with emergent properties is the impossibility to predict their development in the long run because of the nonlinearity and indeterminism of change, where uncertainty and surprise, abrupt processes of catastrophic and non-catastrophic change happen. Yet, the long-term future of the systems and their development is decisive to act in environmental governance in the sense of sustainable development. How this impossibility of acting for the unknown distant future can be dissolved, knowledge about this future can be generated, is the guiding question in this and all following chapters. For the beginning it suffices to say, that the discussion so far concentrates on alternative ways of envisioning the future, especially through scenario analyses (Schwartz 1991; Swart et al. 2004), but further methods of the exponentially growing future studies (Gabriel 2014; Cevolini 2016; Wilenius 2017; Groves 2017) need to be applied in governance analyses.

The environmental problems that appear today at all levels from the local to the global indicate processes of change in social and ecological systems. These changes are consequences of human modification of nature and ecosystems in the many forms of natural resource use, such as hunting and fishing, agriculture and forestry, animal husbandry, mining, industrial production and urbanisation. The modification of nature in human history includes irreversible changes through environmental pollution, disturbing the functions of ecosystems, changes of ecosystems, and global material cycles; when the planetary boundaries of resource use are exceeded, as is today the case when several boundaries are already known and described (Rockström et al. 2009), the problems reinforce and environmental change accelerates.

The diagnosis about the state of the earth system is described in the natural-scientific assessments of the global environment for the epoch called Anthropocene (Steffen et al. 2007) that began with industrialisation as strong modification of nature with effects at global levels, accumulating in global environmental change in the twentieth century: anthropogenic climate change, biodiversity reduction, land use change and urbanisation. Steffen et al. divided the anthropocene in three phases of early anthropocene (1800–1945), "great acceleration" (1945–2015) and a third phase after 2015, "stewardship of the earth system", where the transition to a global sustainable society is on the agenda of global environmental governance. The data they provide, the interpretations and conclusions regarding the effects for nature and society can be interpreted differently. The term Anthropocene is contested, because it refers to humankind as a collective subject, whereas the subject of environmental change is the modern society and economy, not all humans pollute the environment in the same forms and degrees. This epoch of global environmental change implies a strong modification of nature through humans; whether this means an amalgamation of nature and society and dissolution of this dualism or not, remains controversial (see Box 2.2).

Assessments of the social and environmental consequences of the human use of natural resources and the state of the environment exist in different forms: in periodical forms as the annual "State of the

World" reports of the Worldwatch Institute, and in large assessment projects such as the Millennium Ecosystem Assessment, the reports of the IPCC about climate change and adaptation to it, the IAASTD (2009) report about agriculture. Such studies have been intensively reviewed and discussed; they are not discussed in detail here, but used as sources for a social–ecological discussion of global change processes. For this purpose, the knowledge problems in global assessments need to be identified. Such assessments require the analysis, review and integration of enormous amounts of knowledge from different scientific disciplines and specialised fields of research which is always done selectively. An intensifying discussion of interdisciplinary knowledge production since more than fifty years, meanwhile rather advanced as the discourses of "the new knowledge production" and "transdisciplinarity" show, has not yet created advanced methods for generating knowledge and analysing global data sets from natural and social-scientific research, except for specific purposes such global scenarios or modelling of global change processes. Much of the interdisciplinary knowledge production and integration happens in preliminary, methodologically badly developed forms of (see further discussion in Chapters 5 and 6).

Global environmental governance requires for its further development theory-guided analyses of the historical modifications of nature through human labour, and the changing relations between science, economy and politics in the history of human societies. The social–ecological knowledge about environmental change that is so far available is summarised in the following parts for environmental change in historical societies and for present global environmental change, showing the historical variations of human modifications of nature, the dynamics of human societies in their long-term development, and of changing human relations with nature in different cultures and civilisations. This knowledge shows the forms and limits of natural resource use as changing from society to society, connected with the historically changing forms of work and production, consumption and lifestyles, combinations of material and energy resources, modifications of ecosystems and development of human-dominated ecosystems.

The comparison of different historical societies helps to explain the specificities of human resource use in modern industrial society.

Historical comparisons of modes of production, economic systems and human labour are confronted with conceptual and methodological difficulties as the examples of Sahlin's hypothesis of the original affluent society of Neolithic hunters and gatherers, or the more recent anthropological analyses of Diamond about "overshoot and collapse" of human societies show. As doubtful as it seems to describe Neolithic cultures as affluent societies in a terminology from modern society, ignoring their historical specificities, as doubtful seems to conclude from small, isolated island societies in history to large-scale and modern societies regarding their social and ecological limits of natural resource use. Comparative analyses cannot be done without further theoretical analyses of the specificity and path-dependence of societies, beginning in this chapter with a critical review of the changing concepts and views of nature and nature–society relations, to discuss the social and natural causes and long-term consequences of historical and present environmental change. In the following chapters, further aspects and problems of knowledge production and integration for global change and global environmental governance are taken up in epistemological, theoretical and methodological reflection.

The inexact notions of change and dynamics are in this and the further chapters only used for describing more specific forms of change, specified as global social change (discussed further by Goodwin 2009) and global environmental change (Millennium Ecosystem Assessment 2005). Both kinds of processes connect with the more specific processes which are the topic of this book: sustainable development, societal or social–ecological transition or transformation, environmental governance and regulation.

2.2 Human Modification of Nature, Its Forms and Consequences

The story to read in "Science Daily" (November 4, 2013) of the soil as mankind's last resource says, that civilisations rise and fall on the quality of their soil. Loss of fertile soils, droughts and climate change count high among the multiple reasons for the collapse of agricultural civilisations and societies in human history. The reasoning is convincing insofar as loss of fertile soils through pollution, erosion and desertification counts among the multiple reasons and causes of wrong use, overuse of natural resources, and human modification of nature with long-term consequences. But there is more to understand than the quality or loss of soils to explain the collapse of socio-metabolic regimes; soils are used in complex processes of interaction of social and ecological systems, together with other natural resources and in combined use of social and natural resources. Resource use cannot be reduced to quantities and technologies, ignoring cultural-anthropological, historical and sociological knowledge about environmental problems. Such naturalistic reductionism, existing in various forms in environmental research (see below), results in misinterpretations of the kind that all civilisations and societies in history rise and fall for similar reasons, mainly environmental problems.

In a social–ecological perspective, using interdisciplinary knowledge from environmental research, the human and societal relations with nature appear as changing throughout history. In each form of society anthropogenic changes of nature that show a complex causality of human influences on nature with interacting

- symbolic components (worldviews, religions, social traditions, beliefs, scientific and practical knowledge) and
- material components (technological inventions, use of technologies; socially organised labour and modes of production; social, economic and biological reproduction).

In these processes causal relations cannot be simplified to a social dynamic of rise and fall, overshoot and collapse, where political power, building of empires and states, conquering and warfare seem to become the explanatory factors together with biological factors. With the analysis of historically specific modes of production and socio-metabolic regimes of natural resource use it is possible to explain the forms of evolution, adaptation, change, transition or transformation of a society more precisely. Not all societal transformations in human history can be said to be the result of collapses of societies.

The following historical description of human modifications of nature is not detailed, uses exemplary social–ecological analyses of large-scale societal complexes, civilisations or societies (Fischer-Kowalski et al. 1997): early societies of hunters and gatherers, agricultural civilisations and modern industrial society as global society. The chronological account of ancient, medieval and modern times and societies is not used in this description; its reductionism is that of a euro-centric perspective of Western history and society which reduces history to the formative stages of European civilisation and its "ecological imperialism" (Crosby 1986).

Box 2.1 Early human cultures in social–ecological perspective

1. *Societies of hunters and gatherers* dominated for the longest part of human history in successful adaptation to the environment. The resource use in these local, mobile societies remained at low levels of intensity and modification of nature. Hunter and gatherer societies that survived longest in human history, the African, Australian and Amerindian, ended in modern society, during and after the European colonialism. Levi-Strauss saw all of them doomed to vanish under the influence of modern society, less through adaptation and integration, more through violence and extinction. The early societies and their social metabolism help to understand the long-term trends of intensification of natural resource use that mark the pathway of human civilisation, in modernity called "development". Comparing the metabolic profiles of societies or modes of production in ecological terms— hunting, fishing, gathering, nomadism, agriculture, handicraft and industry—shows that intensification was driven through the interplay of several factors: local resource scarcity, population growth and technological innovations that achieved their combined dynamic effects within the agricultural societies and through further cultural differentiation and development of resource use practices, as described by Boserup (1965). An ecological explanation for the transition to new modes of production was provided by Georgescu-Roegen as transformation through "Promethean revolutions", revolutions in energy use from which only two were relevant in the *"longue durée"* of human civilisation: the Neolithic revolution as transition to agriculture and the industrial revolution in modern society as transition from bioenergy to fossil energy sources. Hunting and gathering appears, in all its forms across the globe, with the explanations of Boserup and Georgescu-Roegen, as mode of production with slow change, strongly integrated in nature (uncontrolled use of solar energy, material resources mainly in form of biomass use). The breakthrough on the development path towards modern society and resource use intensification happened

with the Neolithic revolution that brought agriculture, settlement, and the invention of the city. The annual per capita-use of energetic and material resources in agricultural societies was about four to five times higher than that of the earlier hunter and gatherer societies (Fischer-Kowalski et al. 1997).

2. *The agricultural civilisations since the last 10,000 years of human history* vanished, only relicts of them exist today, as for earlier hunter and gatherer modes of production. The agricultural civilisation with the longest continuous existence in human history is that of China, existing about 4000 years with a similar peasant mode of production, isolated from other civilisations. Its rapid transformations happened in the twentieth century. Other agricultural societies in Europe, Asia, Africa and the Americas, the first large-scale political and societal systems or world systems in human history, collapsed for a variety of reasons among which ecological ones, overuse of their natural resource base and ruin of fertile agricultural land were important, as described in cultural anthropology, human and social ecology.

A main consequence of agriculture is forest clearing and the opening of fields for ploughing and planting, combined with animal husbandry. With the Neolithic revolution and the spreading of agriculture in different parts of the world deforestation became a significant environmental change and lasts since then, with a reduction of the forest mantle on the earth from about 40% of the land cover to 20% in the last thousand years. Investigating deforestation more intensively shows a differentiated picture with changing patterns of deforestation and regrowth in human history, as Williams (2000) has summarised for deforestation since the early societies. Knowledge about forms and size of deforestation in early human history is insufficient; much of paleo-botanical and archaeological research is based on modelling. But also for present deforestation of tropical rainforests it is difficult to obtain exact data. The dynamics of forest use, deforestation and replantation, change with different forms of resource and land use and modes of production, differences of climate and types of forests. Yet, in the long-term perspective of human societies, deforestation appears as a main factor of environmental change that became finally one of global dimensions and effects.

Sources own compilation; sources mentioned in the text

The agricultural civilisations included different types of large-scale societies or historical world systems:

- the ancient Near Eastern and Mediterranean societies: Sumer, Phoenicia, Egypt, Greece, the Roman Empire;

- the Asian world systems, especially China and India, with a specific Asian mode of production (Wittfogel) and oriental despotism;
- the Amerindian societies (Maya, Inka).

In most of them—with the exception of China—ruin of fertile soils was one of the factors accelerating their final collapse. Other important factors, especially the conquest through other states or empires were historically relevant until the long rise of the modern Western world system, the prehistory of which for the years 900–1700 is described in environmental history by Crosby (1986) with the hypothesis of an ecological imperialism of European societies. Two forms of occupation of new territories were differentiated in this analysis: that of the colonialism in the tropical zone, where Europeans did not settle in significant numbers, and of the colonialism in the moderate climate zones of North America and Australia, where Europeans settled in large numbers, expulsing, repressing or killing large parts of the native populations. Africa remained a specific continent with Western colonialism conquering the territories only late, in the nineteenth century, when all other territories in the global south were already colonised. In Latin America, the forms of Spanish and Portuguese colonialism differed culturally from the Anglo-Saxon colonialism as described for Mexico by Paz (1961). Colonialism, conquering and occupation of foreign territories became a mechanism of unequal development in all historical and the modern capitalist world system. Societal development happened throughout human history through social division of labour and inequality: economic processes of appropriation, accumulation and surplus production in the centres of the world systems based to a large degree on the exploitation of human labour and natural resources in the colonies. Colonisation can be seen as a socio-economic process of intensifying natural resource use and unequal development in the *longue durée* of human history, accelerated with the Promethean revolutions of agriculture and industry. Industrialisation happened historically in a time when the world was conquered and dominated by the modern capitalist world system. This is the first world system in history that reached global dimensions and has no further reserves for spatial expansion and intensifying of resource use; this is a main reason for its present development crisis that resulted in the idea of sustainable development as rupture of further intensification of natural resource use and search for new forms of global redistribution and sharing of natural resources.

European feudalism and the medieval deforestation and wood crisis (Darby 1956) have been described as special cases of development of agricultural civilisations that brought the breakthrough of industrialisation; it became possible through the exploitation of new energy resources of the "subterranean forests" of coal, oil and gas, the fossil energy sources of the modern industrial socio-metabolic regimes. For the human societies of the past, the forms and consequences of natural resource use in social organised forms of modes of production or socio-metabolic regimes sum up to the messages: Social division of labour and social differentiation of classes formed the social and economic inequality and became also the mechanisms of development and modernisation of society and economy, agriculture and industry in the modern capitalist world system. In the long globalisation process since the sixteenth century other global factors become influential in creating further inequality and unequal access to natural resources, that of exponential global population growth. Whereas social and environmental change in earlier societies happened at local and regional levels, the modern world systems brought as new scale of development and of environmental problems that of global economic and environmental change which determines future possibilities and pathways of development. Environmental problems, deforestation, erosion, desertification, pollution and overuse of natural resources happen now at the level of global change.

Modern industrial society is the shortest phase of human history, the last 250 years, with quantitatively and qualitatively dramatic changes in society and modification of nature through humans. In a somewhat simplified form the modern society can be described in two phases: from about 1500 to 1750 the building of the modern capitalist world system through global expansion of Western societies, still agricultural societies; thereafter, until today, the phase of industrialisation where the agricultural modes of production were transformed into parts of the industrial system and the capitalist economy. Peasant modes of production, similar in many cultures as simple forms of commodity production (Tschajanow 1923), become historical and cultural relicts. Yet, still today nearly half of the global population is living in non-industrialised countries and from small-scale agriculture, another dis-simultaneity and inequality of modernisation that adds to the further development problems of modern society.

Industrial society, with specific forms of natural resource use, social division of labour, with modern science and technology, is now the dominant part of the capitalist mode of production that began with agricultural capitalism, in large-scale forms of feudal property and production in the European countries and in the colonial agriculture. The latter can be seen for a certain period as the driving form of modernisation and capitalisation of agriculture (large-scale land ownership, specialisation and monocultures or "cash crop production"). The previous accumulation as the theoretical explanation of the development modern industrial production in classical political economy happened in different forms in the European metropoles and in the colonial periphery: In Europe it was the long, conflicting and complicated process of the expropriation of the original producers, peasants and land owners, dependent from the cultural traditions and forms of land ownership and agriculture; exceptional was the continuity of a peasant mode of production that was only partially and gradually integrated in the capitalist economic system, finally in the second half of the twentieth century when large numbers of small rural producers vanished through technical and economic modernisation of agriculture to capital-intensive production. In the colonies the expropriation processes were simpler—but not less violent—as processes of occupying new territories and deforestation in areas where few people lived, appropriating the "empty land" for modern agriculture.

The industrial society as modern society of the West cannot be studied without its larger global structure of the capitalist economy and world system, but it is specific in several aspects that are of importance for the human relations with nature: With industrialisation began the forms of exponential economic growth and global environmental change that create the present social and environmental problems. In environmental history, this has been described as the extraordinary twentieth century, where the growth of the global economy amounted to 2000%, that of the human population to 400%, that of natural resource use 800%. With that the debates about "limits to growth" and "planetary boundaries" of natural resource use came on the agendas of public policies.

The century of exponential growth seems to mark a peak in the "*longue durée*" of human societies, where economic growth based on unequal exchange and appropriation of natural resources reached its peak. The combined forms of growth in the twentieth century—economic growth, population growth, growth of use of many natural resources, growth of environmental pollution—cannot continue in the twenty-first century and in the longer run; but it is not yet visible how transformations towards a non-growing economy are possible. The growth phenomena are of ambivalent kind. Economic growth is widely seen as the path to wealth, welfare and human wellbeing, although it operates on the basis of dividing society in rich and poor. Population growth is often seen as a reason for the overuse of resources and destruction of the environment, sometimes resulting in the reasoning the large part of the absolutely poor human population in the countries of the global south cause the environmental problems. Yet, it is evident that economic growth is a structural mechanism of the industrial capitalist system; the unequal exchange and consumption of natural resources shows that the largest part of the global resources is used and consumed by the minority of the population of the affluent countries in the global north. Population growth is not a single and coherent phenomenon, has different forms and reasons. Use and overuse of natural resources in the magnitude causing present global change, exceeding the global carrying capacity of ecosystems, is connected with other forms of change, climate change and loss of biodiversity being the most dramatic changes in a short historical time. Disasters and catastrophes, extinction rates of species, erosion, desertification, forest loss, diseases connected with environmental pollution and destruction, are growing exponentially.

These observations and facts do not tell much about the causal connections and the social and ecological consequences. One cannot identify from the observed phenomena of growth the complex causality, the many effects and the feedback between the processes of change. Environmental policy and governance cannot be derived from such abstract phenomena of growth; for the design of governance strategies and regulation practices, the scientific reasoning with the help of mathematics, logics or cybernetics, in quantitative terms and forms of resource use, is not sufficient. Environmental strategies and action programmes require detailed and in-depth system analyses of the causes and consequences of change, of the interaction between social and ecological systems, of the historical and cultural specificity of the processes. Interdisciplinary research and analyses cannot identify quickly solutions in simple forms of "panaceas" that turned out to be ineffective; the methods of analysing complex adaptive systems requiring further refinement. The methodological problem of analyses of complex systems is discussed as holism "*ex ante*", where the complexity is taken as given before the analysis, but only insufficiently reconstructed through research; it should be developed further though a methodological holism "*ex post*", where the theoretical system analysis informs empirical research. In the situation given, with the interaction of global social and ecological systems, the complexity cannot be simply measured, modelled or assessed; the multiple interacting causalities need to be identified and explained, and brought together in coherent analyses or models, which requires a theory of nature–society interaction of the kind discussed in this book. Only through such complicated, interdisciplinary analyses and syntheses of knowledge it is possible to create new knowledge for policy and governance processes.

The anthropogenic changes of nature achieve in modern society another quality than the hitherto known incremental changes and modifications of nature. Environmental problems "explode"

- when more and more planetary boundaries of resource use are transgressed;
- when the global carrying capacity of ecosystems is exceeded and human resource use is beyond the annual production capacity of ecosystems;
- when the human colonisation of nature reaches the genetic modification food products, of plants, animals, humans;
- when global environmental change reaches hitherto unprecedented forms and quantities of biodiversity reduction and climate change.

In the course of human history nature or ecosystems were continually modified through human activity in forms of local overhunting of animals, deforestation and clearing of land for agriculture and settlement, destruction of fertile soils and erosion. Although limited in terms of social and ecological consequences, such modifications of nature have caused catastrophes in historical societies and their decline. In difference to the historical environmental catastrophes that happened at local and regional levels, within the territorial limits of historical world systems, empires and civilisations, the modern economic world system emerging with the rise of modern capitalism has two distinctive qualities: It is not politically unified, and it stretches over the whole globe. The qualitative differences between the historical and present anthropogenic change of nature can be described in a variety of theoretical concepts and indicators (Bruckmeier 2013: 137ff): carrying capacity, ecological footprints, material and energy flow accounting, human appropriation of net primary production of ecosystems (HANPP), planetary boundaries. All of them show, in one form or the other, how the natural resource use of humans reaches global limits or exceeds them. Yet, it is methodologically difficult to integrate in such theoretical models culturally specific forms of natural resource use.

2.3   Theoretical Analyses of Historical and Present Changes of Nature

Beyond the empirical description of civilisations and societies that rise and fall with their population growth and the use of land and soils, a more complex history of forms of interaction between humans and nature becomes visible with the historical descriptions of changing modes of production and forms of societal metabolism. Processes of modification of nature such as agricultural land use and deforestation differ strongly in their forms, size and consequences between the historical societies:

- Deforestation had more local effects of accelerated erosion and desertification of fertile land in the early empires. In Europe the effects reached large dimensions through the feudal society of the middle ages; the internal forms of colonisation and appropriation of land for agriculture reached critical dimensions towards the end of the middle ages, when wood, the dominant energy resource in the agricultural society, and the "forest crisis" triggered the search for new energy resources that ended in the building of the modern industrial society. The first remedies to combat deforestation systematically developed in early modernity, when reforestation and forms of ecologically rational forest use (not to cut more wood than is re-growing annually) were discussed. This happened with the development of modern forest economy, documented for example, in the treatise of John Evelyn ("Sylva, or Discourse on Forest Trees", 1664) and the early scientific textbook of forest economy by Carlowitz, "*Sylvicultura Oeconomica*" (1713), which can be seen as part of the history of the idea of sustainability or sustainable use of forest resources. Deforestation in modern industrial society is a complex process with a variety of forms and reasons described in environmental history; for the late deforestation combined explanations include the economic processes with manifold forms of use of wood, and the ecological processes in which the consequences of natural resource use sum up to processes of global environmental change.

- Land use change had long time in human history the dominant form of opening the land for agriculture and creating "cultural landscapes" that were predominantly agricultural landscapes, shaping the rural landscapes in most countries of Europe until today. Some of them are transformed in landscape museums that are maintained also when no longer agricultural production happens in them. Other parts of agricultural land are transformed through new forms of bioenergy production that were not happening in agriculture before, where the dominant form of production was that of food for humans and farm animals. Now develops on agricultural land a competition between food and bioenergy production.

- Land use change through settlement and urbanisation has been for the longest part of human history limited, only small parts of the land were built land. With the rapid population growth in the twentieth century developed for the first time "mega-cities" with many millions of habitants; such forms of settlement cannot be discussed any longer with the traditional concept of the city, which is an old historical phenomenon, developing with the Neolithic revolution that brought agriculture and permanent settlement as dominant forms of production and living. Modern mega-cities spread invisibly over large areas and their administrative boundaries; according to their resource use measured in terms of ecological footprints, they require land of a much larger size than their territory; cities dominate also urban areas economically and socially through the forms of mobility, commuting and recreational land use of urban dwellers, phenomena that can be called "colonisation of the countryside through cities". The urbanisation of the countryside includes in European and industrialised countries new industrial and economic forms of land use in rural areas (secondary homes, bioenergy production, infrastructures and transport systems for urban resource use, waste deposits).

2.3.1   A Naturalistic Picture of Collapses of Civilisations

From the history of natural resource use problems in human societies of the past as described above derives a reductionist explanation of societal change, appearing in Malthusian thinking. Its main message is that of an unavoidable collapse of civilisations and societies that tend to overuse their natural resource base through population growth, with the consequence of catastrophic collapses after which societal development starts anew, either at lower levels or in other modes of production. A recent example of such a naturalistic view of society provides as concluding argument: "Virtually every past civilization has eventually undergone collapse, a loss of socio-political-economic complexity, usually accompanied by a dramatic decline in population" (Ehrlich and Ehrlich 2013: 1). The combination of natural-scientific, biological and ecological knowledge with a selective historical view of human history in terms of population growth and natural resource use, without analysing the specific societal structures and processes of production and societal metabolism, makes the theoretical analysis abstract and selective. It builds on a series of observations and facts that cannot be denied, yet, the combination of these knowledge pieces gives rise to doubts about the interpretations. The simple form of Malthus (1798) original hypothesis of geometrical growth of subsistence and arithmetical growth of population that creates "naturally" scarcity of natural resources, misery and poverty, is no longer used in Neo-Malthusianism where the forms and phenomena of global resource use are calculated more systematically and exactly, not only for food resources. However, the idea that population growth results in overuse of natural resources is still influential as the reasoning of Ehrlich shows in a naturalistic or biological view of society, with a doubtful form of selective knowledge integration.

To come from the naturalistic account and biological reductionism of societal development to a more interdisciplinary perspective it is not only necessary to combine the biological perspective with theoretical analyses in the social sciences: that would result in further speculative patterns of thinking which ignore the different forms and mechanisms of biological and societal (including economic and sociocultural) reproduction. The basic assumption of biological reductionism, the naturalistic fallacy, is that the paramount reality of nature with eternally valid laws determines the ways of life of humans and the development of societies; humans cannot change but need to accept the basic laws of nature and the embeddedness of society in nature. As more often in scientific controversies, the differences are not that of either ore, right or wrong, but of how the facts and observations are interpreted with the use of theoretical arguments. One can still argue for the embeddedness of society in nature according to the "new ecological paradigm" (Catton and Dunlap), but develop a more differentiated and refined view of this embeddedness or dependence, which would, for example, include different forms and degrees of coupling of social and ecological systems (Bruckmeier 2016).

The first step of overcoming false dichotomies of nature and society or culture is, to reconstruct the multifaceted term of nature in its varying forms. The conclusion from that is, that the dependence of humans from nature with regard to their biological reproduction and subsistence needs to be reconstructed in historically specific forms, cannot be shown through laws of nature that are independent from human and societal influence. In the controversy about Malthusianism, whether Malthus' hypothesis is valid for pre-industrial, agricultural societies, the analyses of Boserup (1965) showed a different picture based on the hypothesis: When the growth of food supply comes close to the carrying capacity of ecosystems, there happened new technological improvements and inventions that helped to increase food production and consequently allow for further population growth. Thus the argument unfolds that carrying capacity is not of static quality, but changes throughout the history of human societies, with new inventions in food production, in short form: "necessity is the mother of invention". If this can be verified for earlier agricultural societies, it is still more valid for modern society, where agricultural chemistry has since the nineteenth century (Liebig) and especially with modern techniques of genetic modification of plants and animals reached hitherto unknown possibilities to increase food production through intensification of agriculture. But also with that final limits to growth and planetary boundaries of resource use exist that cannot be transgressed without negative environmental consequences.

2.3.2 Human Modification of Nature Today—Global Environmental Change

An interdisciplinary ecological theory of human modification of nature through natural resource use and adaptation to new resource problems, with similarities to Boserup's theory, was formulated by Harrison (1993). He reflected the connections between population growth, environmental problems and sustainable resource management in the Global South. The analysis is framed in the diagnosis of a rise of in environmental problems from national and regional to global scale and scope since the 1970s, with the diagnoses of acid rain, ozone hole, and global warming. The theory is not systematically elaborated, more a generalisation from observations about the historical changes and transitions that occurred since the beginning of industrialisation; it is formulated in subsequent stages of natural resource and waste management (Harrison 1993: 323, 246ff):

- Shifts in resource management happen in parallel with changes in ownership and control of resources: As long as population numbers are low and resources abundant there is no need to control resource use and waste, this is a phase of open access to natural resources; when populations grow and resources are used more intensively, controls and attribution of property rights in form of (local) common property begin; when resources become more scarce they tend to be privatised, owned in private property. Yet, the modern form of private property cannot be realised for all natural resources: The air, the water and sinks cannot be privatised, therefore, they are politically managed and subjected to ever increasing regulation and control, widening from local, national, regional, to international scale.

- Resource and waste management phases are to some extent parallel: Gathering goes together with scattering of waste; mining and industrial production go together with dumping of waste, which results in environmental crises and transition, searching ways towards sustainable resource and waste management.

- Population growth as a main factor driving technological and social change is interacting with other changes. The changes in resource and waste management with the introduction of property rights, regulation and control, can be understood as attempts to deal with resource scarcity and environmental problems. Even when the environmental problems are perceived and their causes understood follows no immediate adaptation or efforts of problem-solving. The development process goes through repeated blockages and crises. Technologies and new forms of production and social organisation need to be developed, which requires innovations and time to introduce and apply new forms of production and technologies of resource use. The transition from agricultural to industrial society included a long transition of the energy regime from wood-based and renewable resources to fossil and non-renewable resources.

- Population, consumption and technology affect the environment directly, but further factors need to be included in the theoretical analysis—poverty, inequality, property rights, human rights, political organisation and democracy, economic organisation and markets—to explain forms of societal development and adaptation measures.

Harrisons theoretical generalisations do not give a complete picture of global resource use, the analysis is simplified, but, together with meanwhile existing other theories (discussed in further chapters), it is possible to formulate a social–ecological theory of nature–society interaction. Analysing the factors influencing societal adaptation and change, Harrison (1993: 327) concludes that a third revolution is necessary to achieve within a half century the transition towards sustainable resource use on the globe, otherwise the resource use and pollution crisis will become unmanageable.

Further theoretical codification and interpretation of global environmental change happen in (1) social-scientific analyses of modern capitalism and its forms of modification of nature (Dickens 1996), and (2) in natural-scientific interpretations of modern society.

1. In *modern capitalism and industrial society* the human relations with nature are increasingly influenced and mediated by scientific, especially natural-scientific knowledge. Modern society differs from earlier societies with regard to the symbolic and material interactions with nature, reflected by Dickens (1996) with the question: How does the specialised and fragmented knowledge of nature in modern society mediate the material relationship of humans to nature? In modern society unfolds the complexity of symbolic and material relations to nature that cannot be understood from empirical research and history writing alone, requires theoretical reconstruction, which shows the dilemmas of environmental action and governance. The conclusions of Dickens, going beyond the constructivist view and analysis of nature–society relations in the social theory of nature by Eder (1996), where naturalistic and constructivist approaches remain as opposite and unmediated, are based on a systematic analysis of the social transformation of nature in which material factors of labour and division of labour interact with symbolic factors of knowledge and belief; furthermore, the unintended consequences of social action in form of alienation and the resulting attempts to react to it in social movements, and new ideas of social emancipation, make the interaction between society and nature more complex. Dickens recapitulates the theoretical programme of Marx analyses of capitalism, showing the necessity for interdisciplinary research and knowledge integration from the social and natural sciences in ecological research. Such a critical theory requires further analysis of science that is itself

subjected to the mechanisms and constraints of capitalist society and of power relations that infiltrate scientific knowledge in its production and application. Scientific knowledge can no longer be seen as the objective, neutral and interest-free knowledge form that guides the self-perception of scientists. A more complete theoretical analysis than that of Dickens develops within the theory of capitalist world ecology, complementing the world system theory of Wallerstein.

2. *Natural–scientific interpretations of modern society*, for example, in the deterministic approaches in nineteenth-century evolutionism, in geography and anthropology, or in recent environmental research with the term "anthropocene" (Ehlers and Krafft 2006), remain controversial because of their neglect of socially and culturally differentiated modifications of nature that do not follow natural laws but human and societal interests. Evolutionism and similar naturalistic theories continue in renewed forms. Among these the concept of the anthropocene has rapidly gained influence in the past decade, not yet providing a new theory of society, but a new concept for analysing specific environmental problems in the last phase of modern society that began with industrialisation and its environmental consequences. The anthropocene can be understood as an attempt to reflect the interaction of nature and society from a natural-scientific perspective, complementing social-scientific theorising. The recent environmental history since 1945 is described in this perspective by McNeill and Engelke (2014). It provides a refined description of the historically changing processes of natural resource use without systematic theory connection.

2.4 Human Modification of Nature in Differing Epistemological Perspectives: "Social Construction of Nature" and Changing "Relations Between Nature and Society"

The philosophical and scientific term of nature developed in manifold forms that are usually seen as mirroring the cultural differences between societies and civilisations in forms of worldviews and views of nature.

In a social–ecological analysis and reflection of the term of nature the history of the idea and term of nature is not the main cognitive interest. The history of the idea motivates continuous philosophical, cultural, and scientific discussion about confusing and contradicting forms of the term of nature; as the other big abstraction in the notion of culture it is seen as an abstraction that needs to be de-constructed, showing the historical and cultural variation of the forms of perception and construction of concepts. In a historically specified social ecology the hidden or latent history inscribed in the different codifications of nature should be revealed: different forms of society, natural resource use and different epochs of societal development. General terms for this purpose are "societal relations with nature" and "societal metabolism", a more specific and operational term is that of "socio-metabolic regimes" existing within the historical forms of human societies. Decoding the latent meanings of the concepts of nature that appear in processes of natural resource use, as "natures with societies", is one methodological component of analysing the interaction of society and nature.

In Western philosophy and science, the concept of nature seemed to have lost its semantic diversity of a rich concept with different variants. The development of the modern natural sciences brought special codes of nature within the newly forming disciplines of astronomy, physics, chemistry, biology, culminating in the in the positivist knowledge culture of the natural sciences that made nature into resource deposits that can be described and analysed through scientific observation and research. The philosophically rooted term of nature became outdated, although philosophy of nature continues in some forms, influencing, for example, the ethical discourse about nature and natural resource use. With the dominance of empirical research in the natural sciences and with the discussion on the "death of nature" through mechanical worldviews in early modern science (Merchant 1990), began a scientific conceptualisation of nature that indicates changing societal relations with nature in the process of modernisation: Perception of nature in modernity is strongly influenced by physical knowledge and terminology. It is this knowledge that helps to understand the macroscopic processes of nature in the cosmos, and the microscopic processes of nature at the levels of atoms. Living nature and organisms appear only as a special case within the more general phenomena studied in physics.

On the other side of the disciplinary divide in science, in the social sciences, began another de- and reconstruction of ideas and concepts of nature. It can be seen as a complementary process of creating a historical and social indexing of nature, with relational constructions of nature and society, in theoretical terms as human and societal relations with nature, colonisation of nature, anthropogenic changes of nature. The broader discourse of "social construction of nature" and the more specific theoretical approaches of critical theory, political economy and social or human ecology are connected with these components of social–ecological theory. With natural- and social-scientific processes of re-codifying nature, the philosophical reflection of nature is given up in favour of more theoretically and empirically specified knowledge. The changes can be seen as consequences of the progressing empirical knowledge creation in the natural and environmental sciences.

Since the end of the twentieth century intensifies a debate about the human modification of nature developing with the new forms of genetic modification and engineering of plants, animals and humans. It can be divided into two groups

- the ones that discuss only present changes in a natural-scientific perspective, without going back to compare earlier concepts of nature and such from other cultures and civilisation (the dominant view in genetics that does not see it as a culturally specific phenomenon, simply as a form and application of natural-scientific knowledge), and
- the ones that include such historical or cross-cultural comparison (Elvin 2010, the research in environmental history).

In both cases the motivating reason of the debate is the present change in nature that is perceived as dramatic—either the macroscopic changes discussed as global environmental change, or the microscopic changes through biotechnology and genetic modification.

In the non-scientific ecological discourse, the discussion of concepts of nature is simplified to some basic and abstract concepts such as nature as the non-artificial world or the wild, physical or dead and living nature, evolutionary nature described in biological terms and human nature. The differentiations between anthropo- and bio-centric views of nature, cannot easily be specified in scientific terms or terminologies, both perspectives remain variants of normative philo-sophical terms and of culturally differentiated worldviews, described, for example, in environmental sociology and cultural anthropology. The discussion varies between the different perspectives and assess-ments as "nature has lost its meaning" (Andersen 2015), "new con-cepts of nature", "nature has changed" and "perceptions of nature have changed", in debates without clarification and consensus, only articu-lating different standpoints. More critically discussed in the broader ecological discourse is the traditional metaphoric notion of "balance of nature" which seemed to articulate a harmonistic view of nature.

Snell and McGuire (2016), in an effort to systematise ancient and modern concepts of nature, start from the following observations and questions:

- Only few concepts can compete with the significance of the concept of nature in Western philosophy.
- The cognitive purpose of the concept was to understand how the world is and how it ought to be, also providing understanding of human nature, as well of the purpose of human life and its limits.
- Throughout this history of philosophy, the concept of nature was used to set the norms and standards for human conduct.
- In contemporary debates nature seems to have lost its significance as guiding ethical and political judgements.
- Nature has become meaningless, a blind process of evolution, whereas the cognitive problems have become such that anything goes with nature, it is possible to accept or reject every characteristic of nature and human nature.

The consequences of this diagnosis of the present situation and human condition are manifest in social processes that go without ethical and political deliberation, in technological progress in biotechnology and artificial intelligence, and in economic processes destroying the com-mon good and the environment. Social and political problems as pov-erty, crime, drug use, war, and threats to security appear more and more as engineering problems that can and should be solved once and forever, in socially risky and authoritative processes where science takes the role of making collective decisions for all members of the society to which it is not legitimised, neither morally nor politically and democratically.

Facing such consequences, a series of questions come up: Why did the older conceptions of nature give way to the new—were they outdated and why? Were they superseded by discoveries in the natural sciences? Were they unsecure and illusionary constructions, or errors? Is a revitalisation of premodern ideas or their reformulation in modern terms desirable or possible? The discussion about such questions on the conference in the Agora Institute in Philadelphia in 2015 did not provide sufficient answers, more underlining the differences between ancient, medieval and modern understanding of nature and human relations to nature. The authors end with the following expectation and conclusion, but no definite results:

- a moderate expectation that the discussion of modern views of nature reveals the historical, philosophical and theological sources of present debates of our relationship to nature;
- a moderate conclusion summarised in the statement: The modern world has separated what nature once held together; nature can no longer reconcile or hold together the sacred and the secular, faith and reason, subject and object, grace and nature. At first, it is now necessary to understand better the changing meanings of nature and what they imply for human action.

This can be read as a comment to the "disenchantment" of nature, aiming to close the debate about a "return to nature", and to deal with the consequences of what modern society, science and technology have done with nature through its modification. The characterisation of the debate of changing views of nature by Snell and McGuire can be seen as a paradigmatic example, illustrating the present debates accompanying such scientific events and discussions as the human genome project, genetic modification of life and cloning, anthropogenic changes of ecosystems and the earth system. The changing views of nature seem to show disorientation, but also the limits and the selectivity of a conventional, academic, disciplinary discussion of the problems. The debate oscillates between the search of normative principles that provide orientation for collective action and the acceptance of the disenchantment or de-sacralisation of nature that happen through the modern natural sciences.

Other analyses of the historical and cultural variations of the conceptualisation of nature come up with different temporal and spatial perspectives, different descriptions and interpretation, refining the description of changes, but no strongly differing conclusions. Modernity, culminating in the global social and environmental change experienced during the twentieth century, appears as unprecedented and dramatic change, and the concomitant reflections in terms of changing concepts of nature confirm the loss of metaphysical and philosophical meanings of nature that survive only in personal, individual worldviews. Societal relations with nature are not defined any longer in philosophical, cultural or religious terms in the dominant Western culture in modernity. The relations with nature in modernity show two experiences of nature: an extreme complexity of global interaction of nature and society, and simultaneously a loss of cultural and ecological diversity that counted so far in human history as guarantees for social and ecological safety, supporting the view that nature cannot be destroyed by humans.

The long history of reflections about nature and analyses of nature ended in diagnoses of the vanishing of wild nature and human appropriation of living nature, in which the concept of nature lost its meaning (Andersen 2015). This evoked new questions: (a) how to conceptualise the interrelations between humans, society and nature ontologically, in terms of separate realities or of a unique reality; and (b) how to conceptualise theoretically and historically the phenomena of nature–society interaction in which the two concepts are connected. Simple forms of conceptual merging nature as "socionatures" or "technonatures" contrast with more refined dialectical concepts that describe the interactions in historically specific forms, do not dissolve or level out the conceptual differences, but show the processes of co-production and co-evolution of nature and society.

An example for confusing and contradicting analyses of nature–society interaction because of insufficiently differentiated and specified theoretical concepts is given in Vogel's (1996) review of the concept of nature in critical theory. Vogel attributes the incoherence in the concepts of critical theorists (including Lukacs, Horkheimer, Adorno, Marcuse, Habermas) to ambivalence and inconsistent forms of use and blending of idealist and materialist, naturalistic and constructivist, anthropo- and eco-centric views and concepts of nature which can be dated back to the philosophical origin of the discourse. He uses as guidelines for his analysis the assumption of methodological dualism and the "misapplication thesis". With that he argues that natural-scientific concepts and methods are applicable for the domain of nature, but not for society, where they become reifications, reinforcing the institutions and the ideological justifications of capitalist society.

The review of Vogel shows how difficult it is to deal with inconsistencies in theoretically codified and abstract terms. An example is his differentiation between

- a misleading question "how ought we to interact with nature" (which he interprets as implying a view of nature as independent from humans) and
- an adequate question what should "the communicatively and practically constituted world we inhabit be like" (Vogel 1996: 168)?

With this alternative a complex and multilayered social construction of forms of interaction between nature and society shrinks to an either-or question, blurring various forms of logical, epistemological, ethical and theoretical reasoning. As a consequence more cognitive problems come up than are solved with a rejection of the dualism of nature and society. This dualism is not static, can be dissolved in a series of more specific theoretical concepts and levels of analysis, with meaningful distinctions that show the different quality and otherness of nature and society, and the historically and culturally specific differentiation of the human and social relations with nature. For the ecological discourse in science and politics, meaningful conceptual distinctions between society and nature seem necessary to show the variability of nature–society relations. A dualistic conceptualisation of nature and society is then not a final theoretical construction that needs to be treated as an ontological distinction (in the forms discussed in the history of Western philosophy of nature), but as a multi-scalar concept which can and needs to be differentiated for specific cognitive purposes. This includes also showing the forms of amalgamation of nature and society in material and symbolic terms, the anthropogenic modifications of nature and non-intended consequences of "nature striking back" after such modifications.

In the present environmental sciences can be found simple and complex attempts to conceptualise the relations between nature and society: in most of the natural sciences, in ecology and in interdisciplinary subjects as human ecology, a predominant worldview is that of nature as paramount reality of which humans and society are part, with an inclination to ignore much of the knowledge from social sciences and modern society and the nature-modifying activities of humans. Simultaneously developed other forms of critique of traditional views of nature and society, of dualisms like "nature–culture" or "nature–society", found in the recent discourses of postmodernism, feminism, and political ecology. The controversies are not finally solved; a schism remains between reductionist constructions of nature as the material existence of physical or living systems and relativist constructions with subjective and culturally specific descriptions of nature or its interaction with society.

In some interdisciplinary fields of research developed more complex, theoretically reflected descriptions of varying and historically specific relations between society and nature, based on interdisciplinary theorising. An early example is found in the theory of Marx (the philosophically formulated approach of historical materialism and concepts as societal metabolism and relations with nature), a more recent one is that of Descola. The historicity and historical variations of nature, culture and society in the long coexistence of humans and nature is the salient point in these theoretical reflections that is not easily reproduced in other forms of theorising about nature and society where concepts are simpler, formulated in general and static definitions. More recent examples of critical relational conceptualisations of nature and society can be found in the discourses of social ecology (renewing, specifying and modifying the concepts of societal metabolism, societal relations with nature, and colonisation of nature), and political ecology when modified concepts from older political economy were discussed—accumulation regimes, metabolic regimes, power relations, gender relations, class relations. Such approaches do not necessarily dissolve the distinction between nature and society, although they support, especially in political ecology, the inclination to dissolve theoretically explicit concepts in manifold subjective, often metaphorical constructions of nature, relativizing concepts in a plurality of subjectively constructed worlds.

Concluding from the two examples of conceptualising nature and society, one in the tradition of philosophy of nature (Snell and McGuire 2016), one connecting to the discourse of critical theory (Vogel 1996), it seems possible to argue: Both discussions do not show how to deal with the dualism of nature and society, and how to make the terms applicable in environmental research; rather they are stuck in theoretical controversies. Other forms of dealing with the concepts of nature and society seem possible and useful, for example, the use of various theoretical concepts:

a. general concepts of nature and society as they are used in the philosophical discourse but also in modern environmentalism,
b. theoretical concepts to specify the societal relations with nature and the historical forms of nature–society interaction, and
c. varying concepts and interpretations of nature and society in different cultures and by different social groups.

Historically earlier societies may have been more homogeneous in their social relations with nature, influenced through cultural worldviews and religions, but not in all of them people lived in harmony with nature: In various historical forms of societies and agricultural civilisations environmental destruction and ruin of the natural resource base for subsistence can be found, more as problem of social practices of natural resource use than of cultural ideas and worldviews. To differentiate in theoretical analyses between the concepts of nature, society and societal relations with nature requires further theoretical clarification and specification (Box 2.2).

Box 2.2 Different concepts of nature, society, their interaction, and societal relations with nature

1. *The concepts of nature and society* as distinct spheres of reality, materially or symbolically constituted, can be used irrespective of their different interpretations as ontological, epistemological, theoretical or analytical distinctions, or—simplified—as distinctions of spheres of reality. Through their connection the concepts show the dynamics of the processes developing during the twentieth century as global environmental and social change. To use different concepts in combinations and combined explanations seems to provide a better theoretical accounting for the complexity of processes in coupled social and ecological systems than shortcut constructions of "hybrid" concepts like "socio-natures", "culture-natures", or "techno-natures". With these shortcuts the historical variations of the social–ecological processes dwindle in conceptual mergers that articulate only the observable forms and phenomena, without providing further explanation of the underlying social and ecological processes.

2. *The theoretical concept of "societal relations with nature"*, encompassing as well material as symbolic components, makes sense when it is distinguished from simpler or more diffuse concepts as that of human relations with nature, views of nature, and worldviews or paradigms (e.g. in environmental sociology by Catton and Dunlap: the human exceptionalism paradigm and the new ecological paradigm). The simpler concepts codify mainly normative and symbolic relations with nature, in forms that are not clearly connected with specific social practices of natural resource use in a society, in agriculture or industry. The cultural, normative and symbolic relations of humans with nature show a greater variability and plurality, do not necessarily reflect the systemic quality of societal relations with nature that cluster in historically varying patterns that require theoretical explanation through a theory of society–nature interaction. Symbolic relations can be conceptualised at individual or collective levels, for social, cultural or religious groups, or for certain professional, gender or age groups. In modern society can be found a variety of "freely floating" symbolic relations with nature, also such from other cultures or earlier societies, detached from the societal relations with nature.

3. *The historically specific components of societal relations with* nature can be described when further concepts are used in the analysis of nature–society interaction. For modern industrial society this implies the mediating forms and processes of interaction that are specified at the level of societal systems: the global economic system of capitalism and its accumulation regimes, the socially organised mode of production or the societal metabolism and socio-metabolic regimes. These systemic processes include operational forms of the theoretical conceptualisation of socialisation of nature at the levels of social and societal systems—in science, politics, economy, culture, specifying the historically varying material and symbolic relations with nature and natural resource use existing in societies.

When theoretically differentiated concepts are used and combined in the analysis of society–nature interaction, the confusing and manifold concepts and descriptions of nature, society and their interaction can be analysed more systematically at different levels, and structured through classifications and typologies.

Sources own review, based on the concepts used in social ecology and in this book

The concept of societal relations with nature discussed and differentiated above is the theoretical core concept in reconstructing varying historical interactions and configurations of society and nature in the processes of natural resource use, in biological economic and societal forms of reproduction, searching for explanations of the historically varying forms of natural resource use and its organisation in society. This can also be formulated as showing the multiplicity of historical and social differentiations which are not different subjective constructions, but socially relevant practices of nature–society interaction in human work, including material production and knowledge production. Theorising the "relationality" of society and nature in a set of differentiated concepts seems a way to deal with the multiplicity of nature concepts in the further discussion of anthropogenic modifications of nature and their consequences, in complementary analyses of global social change in modern society and its connections with environmental change. The development of relational concepts, where neither nature nor society are seen simply as different or as merging, but reconstructed in historical forms of interaction, coupling and clustering, opens further possibilities to integrate knowledge from natural and social-scientific environmental research.

2.5 Conclusions: Long-Term Social Consequences of Human Modification of Nature

The first conclusion from the discussion of multiple concepts of nature in the scientific and political ecological discourses is: to interpret the observations about the forms and consequences of anthropogenic modifications of nature with theoretically elaborate and systematised concepts of human–nature relations that can be found in the inter-disciplinary sciences of social, human, cultural and political ecology. Unfolding the relational concepts implies to classify, specify and differentiate them—in three dimensions, with further differentiations:

a. *human relations with nature* in the sense of the biologically conceptualised human as human species and as individual human body or organism (connected with the "Malthusian" perspective of population growth, biological reproduction and metabolism, subsistence, and limits of natural resource use);
b. *social relations with nature* at various analytical levels, for historically differing social groups and classes, economic systems, modes of production, and societies as total social units (in these forms unfold the theoretical terms of "societal relations with nature" and "societal metabolism");
c. *culturally varying social constructions of nature–society relations* in scientific, cultural, religious or spiritual forms that do not coincide with societal relations with nature at the level of interaction of societal and ecological systems, although similarities are possible (at this level exist social interactions in the lifeworld and manifold subjective constructions and conceptualisations of nature and society).

The three levels of conceptual differentiation do not represent three separate ontological realities or spheres of reality, but interconnecting, overlapping and complementary perspectives that help to analyse and classify the multiple forms and changing relations between society and nature in the course of human history. The systematisation and classifications based on it help to deal with the complexity of the multi-scalar processes mediating between nature and society and the resulting modifications of ecosystems through human use of natural resources in terms of work, production, subsistence activities, science and technologies. The knowledge perspective and the knowledge practices described in analyses of nature–society interaction allow them to describe for each historical society or civilisation, the long-term consequences of human modifications of nature that reach beyond the temporal horizons of a specific society, historical epoch, mode of production or socio–cultural evolution: accumulating effects, positive and negative feedbacks, synergies and forms of adaptive and maladaptive change.

The negative, long-term consequences of human use of natural resources in the historically developing forms of society are connected with the modes of production and socio-metabolic regimes:

- For the early human societies of hunters and gatherers the negative consequences of resource use have been studied in cultural and ecological anthropology, supporting the conclusion: the low population density, small size of human communities or local societies and the simple mode of production resulted in forms of overhunting and extinction of certain species (e.g. mammoth), but no significant modification of nature through land use.
- For the agricultural civilisations developing after the Neolithic revolution the consequences of natural resource use have been studied in cultural and ecological anthropology and in further interdisciplinary ecological subjects as human and social ecology, supporting the conclusion: the intensification of energetic and material resource use, the development of agriculture and cities, the deforestation of nature and the ruin of fertile land resulted in first long-term modifications of ecosystems at local and regional levels. Whereas it is not possible to speak of collapses of societies of hunters and gatherers with the transition to agriculture—both modes of subsistence co-existed long time in human history—most of the early agricultural civilisations collapsed after hundreds or a few thousands of years, among other reasons through overuse of natural resources and ruin of fertile land.
- For the modern industrial society, the consequences of natural resource use have been studied in anthropological and ecological research, in political economy and ecology, and in environmental and ecological economics, supporting the conclusion: natural resource use expanded rapidly through the industrial mode of production, with the exponential economic and population growth, and the modification of nature and ecosystems at global levels. In a very short historical time of less than three hundred years global limits or planetary boundaries of natural resource use were reached, which gives reasons to think the next Promethean revolution in terms of a de-intensification of resource use.

The history of human modifications of ecosystems, of land and soils through agriculture, forestry and urbanisation, of water cycles and other material cycles, of species diversity and of the climate, has resulted in irreversible changes and long-lasting damages of ecosystems. The continuing modifications of nature and ecosystems, especially erosion and desertification in the past 10,000 years since the Neolithic revolution with the invention of agriculture, remain for long time in nature and society: the ruin of fertile soils through salinization in Sumerian agriculture rests until today.

References

Andersen, R. (2015, November 30). Nature Has Lost Its Meaning. *The Atlantic.*

Boserup, E. (1965). *The Conditions of Agricultural Growth: The Economics of Agrarian Change Under Population Pressure.* London: Allen & Unwin.

Bruckmeier, K. (2013). *Natural Resource Use and Global Change: New Interdisciplinary Perspectives in Social Ecology.* Basingstoke, UK: Palgrave Macmillan.

Bruckmeier, K. (2016). *Social-Ecological Transformation: Reconnecting Society and Nature.* Houndmills, UK: Palgrave Macmillan.

Cevolini, A. (2016). The Strongness of Weak Signals; Self-Reference and Paradox in Anticipatory Systems. *European Journal of Futures Research, 4,* 4. https://doi.org/10.1007/s40309-016-0085-1.

Crosby, A. S. (1986). *Ecological Imperialism: The Biological Expansion of Europe, 900–1700.* New York: Cambridge University Press.

Darby, H. C. (1956). The Clearing of the Woodland in Europe. In: W. L. Thomas (Ed.), *Man's Role in Changing the Face of the Earth* (pp. 183–216). Chicago: University of Chicago Press.

Dickens, P. (1996). *Reconstructuring Nature: Alienation, Emancipation and the Division of Labor.* London and New York: Routledge.

Eder, K. (1996). *The Social Construction of Nature.* London: Sage.

Ehlers, E., & Krafft, T. (Eds.). (2006). *Earth System Science in the Anthropocene.* Berlin and Heidelberg: Springer.

Ehrlich, P. R., & Ehrlich, A. H. (2013). Can a Collapse of Global Civilization Be Avoided? *Proceedings of the Royal Society, B 280*(20122845). http://dx.doi.org/10.1098/rspb.2012.2845.

Elvin, M. (2010). Concepts of Nature. *New Left Review, 64,* 65–82.

Fischer-Kowalski, M., Haberl, H., Hüttler, W., Payr, H., Schandl, H., Winiwarter, V., et al. (1997). *Gesellschaftlicher Stoffwechsel und Kolonisierung von Natur: Ein Versuch in Sozialer Ökologie.* Amsterdam: G + B Verlag Fakultas.

Gabriel, J. (2014). A Scientific Enquiry into the Future. *European Journal of Futures Research, 15,* 31. https://doi.org/10.1007/s40309-013-0031-4.

Goodwin, R. (2009). *Changing Relations: Achieving Intimacy in a Time of Social Transition.* Cambridge: Cambridge University Press.

Groves, C. (2017). Emptying the Future: On the Environmental Politics of Anticipation. *Futures, 92,* 29–38.

Harrison, P. (1993). *The Third Revolution: Population, Environment and a Sustainable World.* London: Penguin Books.

IAASTD. (2009). *Agriculture at a Crossroads: International Assessment of Agricultural Knowledge, Science and Technology for Development (IAASTD).* Washington, DC: Island Press.

Malthus, T. R. (1798): *An Essay on the Principle of Population.* London: Johnson.

McNeill, J. R., & Engelke, P. (2014). *The Great Acceleration: An Environmental History of the Anthropocene since 1945.* Cambridge, MA: Belknap Press of Harvard University Press.

Merchant, C. (1990 [1980]). *The Death of Nature: Women, Ecology and the Scientific Revolution.* New York: Harper Collins.

Millennium Ecosystem Assessment. (2005). *Ecosystems and Human Well-Being: Findings of the Scenarios Working Group, Millennium Ecosystem Assessment.* Washington, DC: Island Press.

Paz, O. (1961 [1950]). *The Labyrinth of Solitude.* New York: Grove Press.

Rockström, J., Steffen, W., Noone, K., Persson, A., Chapin, F. S., III, Lambin, E. F., et al. (2009). A Safe Operating Space for Humanity. *Nature, 461,* 472–475.

Schwartz, P. (1991). *The Art of the Long View: Planning for the Future in an Uncertain World.* New York, NY: Currency Doubleday.

Snell, R. J., & McGuire, S. F. (Eds.). (2016). *Concepts of Nature: Ancient and Modern.* Lanham, Boulder, New York and London: Lexington Books.

Steffen, W., Crutzen, P. J., & McNeill, J. R. (2007). The Anthropocene: Are Humans Now Overwhelming the Great Forces of Nature. *Ambio, 36,* 614–621.

Swart, R. J., Raskin, P., & Robinson, J. (2004). The Problem of the Future: Sustainability Science and Scenario Analysis. *Global Environmental Change, 14,* 137–146.

Tschajanow, A. W. (1923). *Die Lehre von der bäuerlichen Wirtschaft: Versuch einer Theorie der Familienwirtschaft im Landbau*. Berlin: Parey.

Vogel, S. (1996). *Against Nature: The Concept of Nature in Critical Theory*. Albany, NY: State University of New York Press.

Wilenius, M. (2017). *Patterns of the Future—Understanding the Next Wave of Global Change*. Hackensack, NJ: World Scientific.

Williams, M. (2000). Dark Ages and Dark Areas: Global Deforestation in the Deep Past. *Journal of Historical Geography, 26*(1), 28–46.

3

Social Change: Social Agency and Human Relations with Nature in the Industrial Society

The end of the industrial society approaches as a consequence of the scarce fossil resources and the environmental damages through industrialisation, but attempts to conceptualise the possible society of the future and transition paths remain unclear and controversial after several decades of discussion about sustainable development. The theories of society and the research about negative consequences of environmental change say little about the future society and how to achieve it; the conventional answer is: Through new technologies and technological innovation develops a new economy. Theories of the post-industrial, postmodern, post-capitalist society, or theories that highlight certain aspects of societal development (knowledge, information or network society, global or world society, peer-to-peer society), do not discuss transitions to a future, post-industrial or sustainable society. This discussion about social agency, transformation and the future society is mainly done in the environmental discourse, including scientific and political debates, or in the interdisciplinary discourses of social, human, and political ecology.

© The Author(s) 2019
K. Bruckmeier, *Global Environmental Governance*,
https://doi.org/10.1007/978-3-319-98110-9_3

3.1   The Situation: Social Agency, the Material and the Symbolic World

In earlier societies, the material world of nature was connected with humans not only through work, also through magic, rituals and religion; agency rooted in a spiritual relationship between people and nature. Such connections between nature and humans are no longer effective in modern industrial society, although religions continue to exist in secular and privatised forms in a disenchanted world. In sociology, a classical explanation of modern capitalist society by Weber is arguing with the hypotheses of disenchantment, rationalisation, and bureaucratisation, variants of the grand topics of classical theories, alienation and anomie, which may also be valid for the relations between humans and nature. Many environmental movements see the lost embeddedness of humans in nature as the main reason of environmental destruction in industrial society. Such reasoning ignores the historically specific forms and the changes of human relations with nature in the course of human history, in favour of simplified explanations of human embeddedness in nature that argue more biologically than sociologically about the relations of humans with nature. For modern society the human relations with nature need to be reconstructed in different forms, as the conclusions from chapter two showed, including the material forms connected with work and natural resource use, and the symbolic forms in terms of views of society and nature that have been described as "instrumental rationality" (Habermas), or in more elaborate forms in environmental sociology as dominant western worldview (Catton and Dunlap).

The diagnosis of a loss of cultural or spiritual connections with nature is insufficient as a reason for the progressing destruction of nature, also misleading when it comes to the solution of environmental problems. The normative views can differ strongly, do not necessarily generate coherent explanations, they are based on subjective value orientations and convictions. Also when the explanation is arguing with collective or religious worldviews, nature–society relations are not understood better; these relations are no longer framed exclusively in religious forms and terms, appear as varying and contradicting. The interpretations of the Christian religion and its view of human relations with nature remain controversial. Referring to the Old Testament (Genesis) it can be argued that it includes an anthropocentric view of nature where humans are seen as dominating nature, the plants and the animals, a rationality of domination that results finally in the destruction of nature when it spreads in collective practices of natural resource use. This interpretation is in contrast to other parts of the holy text that can be interpreted as providing a stewardship ethic where humans care for nature and improve it. The loss of embeddedness in nature can be explained as a consequence of disenchantment, enlightenment, secularisation of religion and dominance of scientific rationality or technology, which is Weber's answer. But in the construction of abstract worldviews, all societal development and change in human history appear as loss of contact with nature and dis-embedding of society. In cultural valuation the answer how to solve modern environmental problems seems reduced to that of changing the views of nature again, coming back to similar views as in earlier societies, which is the answer of parts of the environmental movements. Both answers are not sufficient to explain the systemic mechanisms and social organisation of modern society that structure natural resource use in socially specific forms of access to and appropriation of natural resources.

Analyses of changing cultural worldviews and environmental attitudes in modern societies do not sum to a simple ethic of reconciling humans with nature. Trivialised forms of worldviews in the everyday practices of consumption are measured in the social sciences as attitudes, for example, in surveys about the spreading of material and non-material values among the population, dating back to Inglehart's (1977) diagnosis of material values as conflicting with ecological values. Such studies can be interpreted differently; the empirical results do not show how value changes are connected with behaviour changes and transferred into changes of social practices, life and consumption styles. Comparing religions and worldviews, different views of nature in Christian and other religions that are more environment-friendly, is not improving possibilities of reconciliation either. It seems socially and culturally unrealistic to expect that large parts of the human population find finally to religious worldviews that allow them to live in harmony with nature. Whatever reasoning is unfolded in normative thinking and reconstruction of human relations to nature, it remains incomplete, does not explain the large parts of natural resource use that happens through the systemic organisation of the modern economic world system for the functioning of which symbolic relations and values are of limited importance.

The deficits of explanations and interpretations of human relations with nature can be seen as consequence of ignoring the social facts of modern society, or as looking at these relations in similar forms as they existed in earlier societies, where other worldviews dominated, for example, such that differentiate only between humans and nature, without a mediating system of the kind of society. A similar dualism appeared in early modern theories of society, when a fictitious state of nature was constructed where society does not exist. Society was seen as emerging in contractual versions where the state of nature and the state of civilisation are differentiated through the societal contract which comprises contradicting requirements for the establishment of modern society:

- the creation of a peaceful social order to end violence and wars, connected with the explanation of the state;
- the control, suppression and regulation of the biological instincts and passions of humans;
- and the creation of modern private property which is connected with the development of markets.

To understand environmental change and the consequences of environmental disruption in modern society further conceptual distinctions and explanations of human relations with nature are necessary:

- human relations with nature in a more diffuse sense and societal relations with nature in a more specific theoretical sense;
- individual relations with nature and socially structured relations;

- relations in terms of actors, their perceptions and agency capacity, and relations determined or limited by societal systems, system structures and institutionalised processes.

The different levels and forms of explanation need then to be connected in a theory that accounts for individual, social, cultural, institutional and systemic components as they appear in modern society, showing the relations with nature as complex, differentiated, and contradicting. A coherent explanation is not possible without an interdisciplinary theory of modern society and its interaction with nature, in which more systematic and complete explanations develop. In sociology the knowledge about society and nature is limited (Box 3.1).

Box 3.1 Theoretical knowledge about nature–society interaction and its explanation

1. *Theory of society*: after the end of "grand theory" (with the theories of Habermas and Luhmann) developed "one issue"-theories (of risk, information, knowledge, network society) that detach sociological theory from the themes of classical authors - the analysis of the systemic nature of modern society, its contradictions and social inequality. In these theories society is characterised through a dominant problem or social practice of development—to deal with risks, to use knowledge and information, to communicate through social and technical networks (more descriptions than theoretical explanations). The crisis in the theoretical discourse is articulated in controversies about the form of a theory of society, its explanations, and the renewal of the discourse. For the analysis of environmental problems and their solution a theory of society in disciplinary, sociological specialisation is insufficient; sociological knowledge can only gain explanatory capacity in combination with further knowledge. Necessary are interdisciplinary theorising and synthesis of knowledge, as done in social ecology under the names of "interaction of society and nature", "societal relations with nature", "societal metabolism", "colonisation of nature" and "metabolic regimes", all of which indicate the search of a new form of theory resulting from combinations of natural- and social-scientific knowledge. Such knowledge synthesis is also required for practical purposes of developing global environmental governance.
2. *Social practices of collective action*: modern society appears as a "blocked society" that cannot solve its environmental problems. It is less blocked through bureaucratic organisation (to which this formulation of Crozier refers), but through its economic system and the

capitalist mode of production that is programmed for growth, intensification of resource use, and environmental pollution. The market mechanism directs not only the development of the economic system itself, moreover that of other societal systems and the social practices in everyday life, work, production, consumption. In an ideological and normative reasoning the dominance of the modern economic system and the world market is seen as that of the neoliberal project of the "adjective-less market economy" (Radnitzky), with globalisation, deregulation of markets and limiting the regulatory capacity of the state. This societal reform project blocks a transformation of the system, and the change of the growth mechanism: solutions of economic crises are sought through further growth, at the risk of further environmental change and destruction. The reasons for the limited capacity of social and political institutions to enable change towards a socially and environmentally sustainable society through reforms and collective action are not, as usually diagnosed, lack of environmental awareness of citizen, human greed, institutional inertia, policy and market failures, all of them bearing limited explanation capacity. The interaction between modern society and nature needs to be explained through the systemic nature of the capitalist world system; its social–ecological transformation requires specific forms of collective action in terms of transformative capacity that are so far insufficiently known and developed; the lack of transformative agency, shows the presently lacking knowledge that can only be developed gradually, through transdisciplinary practices of knowledge creation and collective learning.

3. *The institutional complex of the economic world system* that is dependent on the mechanism of growth is the systemic mechanism creating environmental destruction; this destruction is generated through the actors that and their action in production and consumption, but it is an emergent property of the capitalist world system, requires the whole system. It is supported through the vested interests of global players and the power of economic corporations, as agents of the system, not in their subjective interests; the growth mechanism requires the whole economic system to function, and support through political systems to maintain the economic system through policies and legislation. Theoretical explanations that do not synthesise the systemic, the collective and the subjective levels of explanation remain incomplete and selective. The systemic mechanism cannot be explained through human greed or incapacity to learn and use knowledge, although the system supports greed insofar as it works as a driver of economic growth under conditions of private property of the means of production. Economic growth as an abstract term provides only a simplified form of theoretical explanation for environmental damages and externalities. A more systematic and complete explanation emerges with the combination of different explanation components: the mode of production; the societal relations with nature; the societal

> metabolism and metabolic regimes for maintaining the economic system; the accumulation regimes explaining the functioning of the economic system; the functional division of labour between the economic and the political systems that support the growth mechanism and the property relations; the intensification of natural resource use and the forms of environmental problems and environmental change; the economic practices of valorisation of natural resources and unequal appropriation of natural and social resources; the development of transformative agency, including forms of regulating nature–society interaction, forms of institutional cooperation and governance practices, forms of collective and transformative action, forms of conflict resolution in the transformation processes and forms of collective learning of the actors.
>
> *Sources* own compilation from the sources discussed in this chapter

The differentiated explanations of human and societal relations with nature are required to explain further the present situation, the lacking capacities of transforming the modern societal system, or of societal agency to create a sustainable society. The lacking agency is understood differently in sociological theories:

1. In the *systems theory of Luhmann* (1986, 1998) the impossibility of environmental sustainability in this system is seen as a consequence of the mechanism of functional differentiation of modern society that has no specialised institutions and subsystems to communicate with nature or dealing with environmental problems. The institutional communication of environmental problems and the internalisation of "social costs of private enterprise" is blocked through the lack of generalised communication media that allow to deal with environmental pollution. Media as power and authority, legislation, money and capital, scientific knowledge, value commitments, are not effective to solve environmental problems. The problems create only resonance and disturbance in the societal system, but no effective solution through governmental decisions or collective action. In this view modern society can only evolve and adapt, but not transform into another type of economy and society.

2. In the *older critical theory* of the Frankfurt School analysis of environmental problems is not sufficiently done with social-scientific and

ecological knowledge, more reflected in philosophical terms where eschatological connotations are implied in the notions of reconciliation with nature or seeking a paradise lost (Alford 1985: 174); with these notions, the mechanisms of the modern societal and economic systems that make human reconciliation with nature impossible, or an utopian hope, are not analysed further. The incomplete reflections show nature as "the other" of society, excluded from social relations, with the consequence that environmental problems are not solved, but disturb the further development of society through non-intended consequences of social action. The *newer critical theory* (Ludwig 2013) is still on the way from a social–philosophical discourse about recognition (Honneth) towards a theory of society that is able to explain the development and change of the capitalist system. In world system theory (Wallerstein) the blocks of a transformation of the modern economic world system are explained with the economic mechanisms of the capitalist world system that is based on profit generation and a negative systems ecology, creating a dysfunctional capitalist world ecology that undermines the further development of society through short-term growth imperatives and destruction of the ecological and the natural resource base of society. Similar explanations are developed in critical political-economic theories.

It is evident that the theories reflect and describe blocking systemic mechanisms in various forms and contribute to the understanding of the dysfunctionality of industrial society and capitalist economy, showing forms of maladaptive change rooted in the systemic mechanisms of modern society as industrial and capitalist society. However, these and other theories of society discussed in this book do not create sufficient knowledge for possible ways of transforming society into an environmentally sound society. How to generate transformative agency, to deinstitutionalise the economic growth mechanism, are theoretically and practically unanswered questions. Not only the theories, also the degrowth discourse and movement developing in the past decade deal more with the symptoms than with the systemic mechanisms of dysfunctionality. The question, *how to create transformative action capacity*, brings all knowledge-generating mechanisms in science and other social knowledge practices to their limits. Ideas of creating transformative literacy, transformative agency and capacity, developing a transformative science, creating knowledge for social–ecological transformation, are not yet providing explanations, only formulating the necessity of further knowledge and creating expectations.

Starting from the results of the discussion in the preceding chapter, the concepts of nature and society need to be reconstructed in a social–ecological perspective as relational concepts to begin with the search of transformative action capacity; then the analysis needs to be developed further in a theory of nature–society relations presently discussed in social ecology (see Chapter 7). For the formulation of such a theory it is necessary to work through the disciplinary research about society and theories of modern society, mainly found in sociology, to show their achievements and limits and how to develop further explanation through combinations of concepts and explanations in an interdisciplinary theory. This is done here with regard to the analyses of modern society and of industrial society. It is discussed how theoretical concepts of society need to be complemented with other concepts to develop an interdisciplinary perspective for theorising nature–society interaction.

The questions, how the social spheres of lifeworld and system, of society at large, are constituted, generated and reproduced are to be answered through sociological theories of society. These theories provide competing explanations, not mainly because of the use of different data and empirical knowledge, but through epistemological and methodological controversies about the theoretical interpretation, combination and generalisation of available knowledge. Furthermore, the cognitive and practical aims of a theory of society, its scientific and practical significance and purposes are controversial: Whether and how it should inform and support attempts to transform society is unclear. For the practical application of scientific, empirical and theoretical knowledge about the relations between society and nature, further forms of knowledge production need to be sought outside the established institutions of science and research (in politics, in social practices, in everyday life). Inter- and transdisciplinary knowledge generation and synthesis are required to support the transformation towards sustainability.

The concept of *modern society* is used in various, complementary theories, developing in reaction to the Parsonian orthodoxy of explaining modern society and its emergence through a specific value system of Western culture; also the relations between society and nature are then mainly reflected and interpreted in terms of cultural values, cultural change, knowledge change, technological innovation, rationalisation, and the development of individual and collective forms of rational action. This happens in abstracting from the practices and consequences of natural resource use and human modification of nature. Economic production appears in Weber's typology of social action as explained through its "instrumental rationality"; the sociological explanation ends with that, not explaining further the systemic nature of the capitalist system that is only described in historical studies and through comparison with other systems. The concept of *industrial society* develops in a more social–ecological perspective where the interaction between society and nature and natural resource use is implicitly dealt with, although not always reflected theoretically in term of nature–society relations. Industrial production is not only a specific form of social action following a specific rationality; it is natural resource use through human labour; the material practices of production and consumption, the appropriation, use and management of natural resources, and the modification of nature through human labour and technologies need to be reconstructed theoretically.

The concepts of modern and industrial society do not cover the whole spectrum of sociological theories of society, and also not show the great divide that characterised sociological theorising throughout the twentieth century, that between traditional theory (constructed according to the epistemic model of the natural sciences) and critical theory (developing from the theory of modern capitalism by Marx). Yet, both concepts become important for the analysis of nature–society interaction; other theories of society discussed in this chapter refer to these two as providing the core of a sociological explanation of modern society that needs to be complemented through more systematic analyses nature–society relations that are neglected in the disciplinary tradition of sociology.

3.2 The Discourse of Modern Society and Its Limits

Before the institutionalisation of an academic discipline of sociology towards the end of the nineteenth century, theories of modern society influential until today developed in various forms in political economy (the theory of modern capitalism), philosophy (theory of the social contract), through influences from the natural sciences (positivism, evolutionary sociology), and in the first half of the twentieth century influences from cultural anthropology. The theoretical discourse changed with the development of academic standards of theorising, where the verification of theories reduced to epistemological and methodological criteria of knowledge generation, giving up the idea to verify a theory through its transfer and application in social knowledge practices for changing or transforming society, politically or in other forms. Most of the theories of society that developed in the twentieth century in sociology reflect only weakly the differences between traditional and critical theory which Horkheimer described as the theoretical in the theoretical analysis of capitalist society. The differences between traditional and critical theorising dissolved in the past decades with the theories of Habermas (adopting in its final form as theory of communicative action functionalist components), Giddens (developing through a critique of Marxist theories and modification of the philosophy of praxis), Beck (developing through a critical analysis of the modernisation process that includes environmental and technological risks). Somewhat surprising is the attempt by Amstutz and Fischer-Lescano (2013) to develop a new form of critical theory of modernity from the systems theory of Luhmann.

   Towards the end of the twentieth century, when environmental problems and economic globalisation became dominant theoretical themes, the sociological theory discourse changed: Grand theories, in traditional or critical variants ended; interpretations of global change resulted in reductionist and preliminary forms of theorising social, economic, political, cultural or technological changes, fragmented in sub-disciplinary and thematic specialisation of research, not synthesised in an encompassing theory of society and nature. The specialised fields of research created fragmented theories of society developing in the past decades as risk, network, information or knowledge society. Systematic attempts to reformulate a theory of society were given up, using the more diffuse term of social theory and a plurality of explanations (Delanty 2006). Attempts to renew older theories (as that of Parsons and of systems theory) or theoretical discourses (as that of critical theory) did not succeed and last, but ended in new controversies.

An exceptional phenomenon in the fragmented sociological discourse is the lasting success of a variant of critical theory, world system theory that survived the end of the grand theory tradition as well as that of the marginalisation of Marxist theories. Regarding the theory of capitalism of Marx the ecological aspects and environmental consequences of capitalist forms of accumulation are meanwhile systematically elaborated in variants of ecological Marxism (Hirsch 1990; Altvater 1992; O'Connor 1994; Foster 2002; Burkett 2006; Foster and Burkett 2016; Chen 2017); also other themes like the internet (Fuchs and Mosco 2016) are interpreted in the framework of Marxist theory. New efforts to develop classical Marxism for analysing present changes in capitalist society have not yet strongly influenced the broader and interdisciplinary discourses of globalisation, remained a specialised critical movement of experts and researchers, gaining influence in other ways: by showing deficits and failures of conventional academic science and the necessity of inter- and transdisciplinary knowledge production, developing in competition to academic sciences.

The intensification of the Marxist discourse about society and nature is, in the development of the broader ecological discourse an indication of the necessity of critical thinking to deal with the environmental problems of modern and digital capitalism. The heterogeneous knowledge cultures of critical theory of capitalism, disciplinary sociology, and interdisciplinary social–ecological theory influence each other in critical debates and controversies. Whereas interdisciplinary research spread continually, interdisciplinary knowledge syntheses, including theoretical syntheses developed slowly, mainly in social ecology, world system theory and critical political economy that developed and broadened their critical system analyses since the beginning of the global environmental discourse in the second half of the twentieth century to include society–nature relations. The social–ecological theory of nature–society interaction integrates and develops the theoretical analyses complex and interacting systems further. Other types of theory in sociology, providing less knowledge and explanation, include the presently developing sociological theories of late modern society (Box 3.2).

Box 3.2 New sociological theories of society and their relevance for analysing nature–society interaction

1. *Modernisation theory and its derivatives*: Classical modernisation theory (Parsons) could be developed further in new and more critical variants of risk society, reflexive and ecological modernisation (Beck, Giddens, Mol) that take up environmental problems in the analysis; this is their important innovation, sometimes together with a neglect of analyses of social inequality as in the classical theories that are seen as outdated. The theoretical reflections and explanations in these theories include the phenomena of systemic risks that can neither be predicted nor controlled through political institutions, the idea of reflexive modernisation that implies the dealing with side-effects and non-intended consequences of social action among which environmental consequences are important, and the idea of institutionalising beyond economic rationality a second form of ecological rationality in social action and decision-making that allows to control and correct the negative consequences of economic rationality. All these new facets of analysis and explanation improve the modernisation concept theoretically without creating a systematic new theory of modern society, rather creating additional explanations to fill the explanatory holes and deficits of modernisation theory.

2. *Information and knowledge society*: Rule and Besen (2008: 141f) analysed the basic ideas and assumptions in the theories of the information society, for example, that of new, especially scientifically based knowledge practices. They conclude that these ideas are not new (dating back to enlightenment thinking), are empirically weakly supported, but resurface repeatedly in the history of modern society. The rapidly increasing information flow through the use of the internet does not yet create a new society that can be said to be the second global system after the economic world system; rather it is a fragment of a new society that comprises so far only the closely linked processes of economic globalisation and technical communication. Obviously the information flows do not create new theoretical explanations of modern or late modern society, but highlight changing social practices of knowledge use that are analysed and described in similar form in the complementary theories of modernisation mentioned above, or in the communication theories mentioned below.

3. *Network society* is represented by the influential theory of Castells, although there are different theoretical applications, for example, in actor–network theory (Callon, Latour) and interdisciplinary network analyses (Watts 2004). Castells (2013: 24f) describes network society as a global society whose social structure is built through communication networks activated by microelectronic, digital information and communication technologies. This is a form of sociological analysis, but whether it is sufficient to create a new theory of a global society can be doubted. In the present early phase of the internet many people are not involved, but everyone is affected by the processes in the global networks that constitute the social organisation in production, consumption, reproduction, in culture and power relations. The internet appears as a technological infrastructure that organises the global society through maintaining and activation of communication networks. The global society became only possible through, and is dependent on this communication technology. The essential process that constitutes the global society is communication, similar as before formulated in the theory of Luhmann, where the internet did not yet appear as the core system, but the global society was explained through global communication. Global communication processes can be described in this way as organised through the internet technology, but this does not explain all societal functions and processes, dos not replace other mechanisms of production and reproduction. It is one facet of modern society, the communication interface and the surface of society; for the explanation of the whole society and its reproduction further structures, processes and mechanisms are required. The global internet-society appears rather as a parallel society, including the parts and processes of late modern society that can be organised through the internet; with the digital divide it produces further forms of exclusion, marginalisation and social inequality. Nature and societal relations with nature can in this communication perspective only be considered as communication about nature and the environment, as consequently done in Luhmann's theory (1986), excluding the material and energetic forms and processes, work and technology that constitute the interaction of nature and society. The reduction of environmental problems to communication about these is also the perspective in which Castells (2013: 299ff) describes environmental problems and global environmental change: as political communication about it in environmental movements and environmental politics: as reprogramming of communication networks. Solution of environmental problems and questions related to that are ignored, or, as in Luhmann's theory seen as impossible with the argument: the modern society with its specific forms of social organisation and functional differentiation has no effective mechanisms to deal with environmental problems.

Sources own compilation; sources mentioned in the text

In *environmental sociology* further theoretical knowledge integration or new forms of theory of society and nature are not yet available. The main theme of environmental sociology, interaction between society and its biophysical environment in different spheres of social action and social systems, is discussed in competing theories from traditional modernisation theory (e.g. Mol's theory of ecological modernisation) and from critical theory of capitalism (e.g. the theory of ecologically unequal exchange by Rice or theories of ecological Marxism).

With the concept of cosmic society Dickens and Ormrod (2007) attempt to describe the relationships of modern society with the universe, the conquering and the colonisation of the cosmos as a futuristic project of humankind, showing the earthly society as one that cannot develop further in the limits of the earth system. Although understood as based on critical theory and connecting to the Marxist discourse (Dickens 1996), projecting the terminology of critical theory of society to the imperial project of conquering the cosmic space in a discussion of the future of capitalism, the analysis suffers from similar deficits and shortcuts as the other sociological theories regarded here. Its selective and limited empirical basis is the possible growth of the space industry as part of the military-industrial complex and of the development of space tourism. Modern society appears in a selective view of its limits of development and growth, among which the limits of the natural resource base on earth are important, although they are not highlighted in the theory that gains its cognitive capacity more form a new name than from new knowledge. The future of society and the solving of present societal crises is discussed in a technological perspective of development, critically reflecting the technical limits of capitalism, and the possible switching of capital to new imperial projects in the outer space in attempts to solve accumulation crises (Dickens and Omrod 2007: x). In times of a long crisis of financial capital and of a crisis of sociological theory of society, the so far hardly developed sociology of the space can also be used to experiment with a renewal of the theory of society; this happens in a scenario of a future society that turns one more time the limits of resource use and growth into a search for new space for expansion. However, the colonisation of other planets is not yet a realistic scenario for the twenty-first century, and the environmental problems need to be solved on this planet.

The following attempt to renew the theory of modern society in a naturalistic theory, mainly using biological knowledge, developed outside the mainstream of sociological thinking as an epistemological renewal of sociological theory (Lenski 1988). The theory connects with the environmental discourse although developing outside the new environmental sociology. Lenski (1988: 163, 170) attempted to connect the theory of society with the emerging environmental theme and interdisciplinary thinking by developing two arguments: A general sociological theory should develop conceptual links to theories in other disciplines, connecting especially to the natural sciences to create more rigorous standards of theory construction and verification; secondly, sociological theory should incorporate biological and environmental constants to analyse the interactions of these with social and cultural variables.

Similarly as in the new environmental sociology of Catton and Dunlap Lenski argued, that social facts need to be complemented by physical and biological facts. His arguments, showing the necessity of transferring knowledge and methods between different disciplines, were not influential in sociological discourse. Lenski attempts to formulate a general sociological theory of evolution to connect biological and sociocultural evolution and specifying their differences. In his approach of macro-sociological and ecological-evolutionary theory (Lenski 2005), he reviews ecological knowledge selectively, although natural-scientific knowledge and theorising are understood as providing the scientific knowledge about material nature, whereas society appears as a secondary, derived and dependent sphere that cannot influence the biological evolution.

With the attempts to formulate a general theory of societal evolution, Lenski eliminates the historical and systemic specificity of modern society from the analysis. The ecologicalevolutionary theory focused on the interaction of genetic and cultural factors in societal development, summarised in a typology of human societies derived from the connection of two influencing factors, that of the natural environment to which a society needs to adapt, and the level of technological development. With these assumptions seven types of societies are differentiated (hunters and gatherers, horticultural, agricultural and industrial societies, fishing, herding and maritime societies). Useful as the classification is for describing ecological specificities of historically existing kinds of societies, it simplifies and reduces the analysis and explanation of differences between societies to a typology of natural resource use. All seven types of society Lenski described are classified in their specific forms of natural resource use and technologies—necessary factors of development, but reductions with regard to the complexity of societies, or the modes of production, socio–metabolic and accumulation regimes of societies. Subsuming of social under biological principles n this theory comes close to older forms of determinist thinking. It shows like in a magnifying glass the deficits of biological thinking about human societies, at times where biological theory was trying to influence social theorising in more successful form through epistemological theory (Maturana and Varela 1980). In social–ecological theory the explanatory capacity develops with a differentiation and specification of social and biological evolution that are connected in manifold and historically varying forms. The analysis of different, contingent and multiple forms of coupling of social and ecological systems can explain better the forms and variations of interaction between nature and society in the course of human history.

Some of the critique of simplified forms of connecting social and ecological knowledge is formulated in the critique of the ecosystem concept by Moran (1990: 15ff) in its applications in anthropology. This is relevant for the interdisciplinary transfer of concepts and knowledge when the interaction between nature and society is analysed more concretely and empirically. Moran identifies a series of critical arguments in the anthropological discourse:

- an inclination to reify the ecosystem concept attributing to ecosystems the properties of biological organisms;
- an overemphasis on predetermined forms of adaptation, for example, using the criterion of energetic "efficiency";
- a tendency of abstracting from historical factors, time and structural change, thus overemphasizing stability of ecosystems;
- a tendency to neglect individuals and their action in favour of populations;
- no clear criteria for defining the boundaries of ecosystems;

- uncontrolled shifting of levels of reasoning between field studies (where only few sites, limited time periods and some aspects of the whole system are studied) and whole ecosystems and macro-ecosystem models.

Several of these critical arguments could be used for a critique of Lenski´s evolutionary theory, especially with regard to its underlying assumptions of constants, stability, timeless structures and primacy of ecological processes. Moran's critique and the more general critique of theoretical and methodological deficits of ecology by Peters (1991) show, that the expectations formulated half a century ago by Baker (1962: 21), when ecological concepts were hardly used in anthropology, have not been realised; they included the expectations that ecology helps to strengthen the analysis of interdependencies between human biology, human culture and the biotic and abiotic environment of humans, and to find more developed causality relations, including multiple and circular causality.

The discussion of theories of modern society shows that the unanswered question is not that of explaining the environmental regulation and governance crisis, but that of finding ways to solve the crisis, which is connected with the question about the future of modern and industrial society. The crisis requires for its solution requires a transformation of the societal system which cannot be conceptualised and initiated without further interdisciplinary knowledge generation and application. Several theories, such from critical theory and the systems theory of Luhmann can explain why modern society cannot solve the environmental problems it causes. When solutions of the global environmental problems are sought, the theories reach their limits: How to change the destructive path of economic growth and what kind of a future society to build is unknown. The first problem can be described as paradox of the industrial growth society that does not end and cannot continue; the second problem as the paradox that the future cannot begin because one does not know how to build a new post-industrial society.

This paradoxical situation shows that the governance crisis is a knowledge crisis of the kind that the limits of scientific knowledge and ways to shift the limits of knowledge in other ways but empirical research need to be discussed. With the interpretation of the crisis as knowledge crisis that limits the development of global environmental governance, it becomes, furthermore, necessary to develop new forms of knowledge communication, transfer, and sharing between science and governance, researcher sand knowledge users. Theoretical analysis of interacting social and ecological systems can provide new knowledge for the initiation and organisation of the transformation process. The relations between theory and governance practice need to be rethought epistemologically. Habermas (1973, 2006) has discussed these relations more generally and with regard to political communication in the media society. From his earlier reflections he draws the conclusion that politics has become, under the influence of science, a technical discipline that lost its connections with the "praxis" of a democratic decision culture and enlightened citizen as participants in politics. Thus he provides arguments for the later developing critical discourse of post-politics, but no more concrete arguments for the development and renewal of environmental research and governance. With his recent discussion of political communication he specifies this diagnosis to develop new forms of democratic decision making—in two complicated conditions: "mediated political communication in the public sphere can facilitate deliberative legitimation processes in complex societies only if a self-regulating media system gains independence from its social environments and if anonymous audiences grant a feedback between an informed elite discourse and a responsive civil society" (Habermas 2006: 411). The answers help to understand somewhat the challenges and requirements to create new forms of democratic governance, but they are not developed and specified for the problems of global environmental governance where still more complex problems of social–ecological transformation are to be solved.

Also in the broader discourse of critical theory the questions are not sufficiently answered that motivated the development of critical theory since the beginning with critical political economy in the nineteenth century; in the new discourse of critical theory is again sought for ways to develop a new form of theory of society (Ludwig 2013). In the continuing controversial debate among critical theorists, about ways out of the intensively discussed production and accumulation crises periodically coming up in the modern economic world system, two solution perspectives clash: perfecting of the system (dealing with the theoretically formulated impossibility of a crisis-free and non-growing capitalism), or transforming the system (dealing with the paradox of building a new society without knowing how to do it). The dilemma is articulated in different forms:

- in the discourse of a critical theory of society (in exemplary from by Gorz 1982) it is discussed as the as changing social subject of societal transformation;
- Gouldner (1970) diagnoses a crisis of the theory of modern society and anticipates the later decline of grand theory;
- in the world system theory the knowledge crisis resurfaces as that of interpreting the present changes in society as globalisation or transition to a new society (Wallerstein): whether the system develops through adaptation or through societal transformation.

Epistemological reflections of these problems of knowledge production and application highlight the construction of rationality in competing theories of society.

Casteneda (1987) shows the older variants of reflecting the knowledge crisis as focussing on the discussion of rationality in the discourse of critical theory, ending with the conclusion that the ways of knowledge production in critical social theory are diverse and confusing, referring to the theories of Habermas and Giddens. The unsettled controversy between positivism and critical theory about how to construct rationality is then discussed further by Cummings (2002), arguing that as well the positivist version as that of Habermas are unintelligible in certain of their claims and self-refutating. The theoretical discourse is according to these critical reflections stuck in the analysis of dilemmas of the rationality of knowledge production and application in social action.

Ortiz (2008) follows up the discussion of objectivity, rationality and normativity of knowledge production, showing the fragmentation and naturalisation of epistemology in dealing with these problems, and arguing for giving up absolute or a priori reasoning in epistemology; he suggests *to reformulate the normative implications of knowledge production in a pluralistic, pragmatic and non-essentialist epistemology.* This seems to be a final conclusion from the long and erratic discussion in epistemology during the twentieth century that has, since Kuhn's (1962) historical and empirical turn of studying scientific knowledge production, drifted away from formulating absolute and universal criteria of scientific knowledge production. What is described and translated in epistemological rules and criteria now are particular norms for specific forms of knowledge production and application, derived from empirically observed scientific knowledge work, with preliminary, temporarily stable and workable criteria for social practices of knowledge generation, transfer and use. The epistemological problems in this knowledge chain from theory and research to practice and knowledge application appearing come up again in the construction of a theory of nature–society interaction that informs the practices of global environmental governance and sustainability transformation (see Chapters 5–7).

3.3 Industrial Society as Blocking Sustainability

In the ecological discourse industrial society appears as a society that does not end and cannot continue, which can explained—to a certain degree—with the hyper-complexity of the modern economic world system and its dual structuring as industrial system and as capitalist system, both with specific functional requirements:

- the industrial system as dependent on the permanent flow of natural resources and their transformation through extraction, production, consumption, resulting in waste and pollution;
- the capitalist economy as a system dependent on economic growth, capital accumulation, valorisation of nature and human labour, and the dominant logic of maximising profit.

The critical statement by Moscovici (1968), that in ecological terms societies are not to be described as capitalist or socialist, but as agricultural or industrial, is simultaneously enlightening and dis-informing.

Enlightening it is, because it shows that there are different social structures and orders coexisting in modern society that are contrasting with regard to the relations between society and nature; dis-informing it is, because it gives a shortcut abstraction and view of the problems to deal with—as if one could separate between the two systemic components of modern society and solve the environmental crisis through a transformation of industrial society without transforming the capitalist system. The cognitive problem can be solved by discussing the forms of abstraction connected with these concepts of economic and societal systems. Capitalist and socialist society are understood in this description more in terms of politically organised and constituted forms of society, less as complex, systemically constituted societies in terms of accumulation, reproduction, and socio–metabolic regimes, a shortcut found in many social theories of modern society. The cognitive problem that Moscovici leaves unsolved is how different forms of social organisation of society interact in modern society. In human ecology and environmental sociology it is discussed as the relations between sociocultural and socio-ecological order and agency in modern society (Manuel-Navarrete and Buzinde 2010). In social and political ecology it is discussed in more systematic and complex form as the interacting components of social–ecological transformation (see Chapter 7). With that it had found the final form in the analysis of the different forms of blocking societal transformation or the perseverance of unsustainable societal systems.

The history of industrial society and its resource use practices are analysed in many disciplines. It is not necessary for the purpose of the discussion of the transformation to a future sustainable society to analyse the societal history again. With the social–ecological knowledge discussed here industrial society can be described and assessed as the shortest phase in human history that has driven natural resource use to final social and natural limits. This has happened in less than three hundred years, beginning in England, spreading in the Western countries, but not in the global south. In comparison to the English industrialisation all other countries appeared as industrialising late, also the United States that industrialised after the civil war from 1861 to 1865. In Europe delayed industrialisation happened in the twentieth century in the socialist countries in Eastern Europe, ending in their collapse that revealed high levels of environmental pollution and destruction of ecosystems. Presently a new wave of delayed industrialisation is on the way in the newly industrialising countries of the BRICS-group. For this process that includes a larger part of the poor global population it is difficult to imagine its completion, as the global limits of resource use are approaching, requiring de-intensification of resource use and reduction of pollution of the environment in future.

The post-industrial society, a premature utopia (Bühl 1983), has not yet begun, only projects of industrialisation in the global south and relocation of industrial production from the core to the periphery of the economic world system. The spreading of industrialisation to the many countries in the global south that are not yet industrialised is, seen in the perspective and with the knowledge provided in social ecology and through other environmental research, finally impossible because it would overshoot the natural resources available on the earth several times. The debate on limits to growth has set this problem on the environmental agenda since the 1970s, but still it is not visible that resource-intensive economic growth is ending, although the planetary boundaries of human resource use can meanwhile be calculated better than in the studies on "limits to growth" mandated by the Club of Rome. The development of governmental environmental policies in many countries and of sustainable development policies mark, so far not successful, attempts to solve the problems without touching the question of transforming the industrial system—in "piecemeal engineering", ecological modernisation of industry and development of "green technologies" that create partial and inconsequent solutions, shifting or externalising the environmental problems. This was described in the theory of risk society by Beck as shifting the burden of environmental pollution to specific social groups as the poor, the women, or to specific countries in the global south, or to future generations.

The historical description of industrialisation happening at regional or national levels, in different forms and at different times, is a simplification of the systemic nature of industrialisation that happened within the modern capitalist world system, in changing forms of international division of labour, with unequal sharing of burdens and benefits. That the Western countries that industrialised first are now achieving a post-industrial stage of development, whereas the countries that industrialise late as the BRICS-countries, achieve that stage later, is not yet saying much about the social and natural limits of resource use that determine the possibilities of industrialisation. Attempts to explain the solution of environmental pollution through industrial production through the Kuznets curve of environmental pollution— economic development and industrialisation are at first causing environmental pollution that is in later phases of development technically and economically solved—are not empirically verified. The project of industrialisation was from the beginning part of the development of the global capitalist economy that showed uneven development and the global divide between North and South, core and periphery, rich and poor countries, throughout its history. The economic mechanisms that made the Western countries industrialising first and the others not or delayed are the same, all are part of and influenced by the capitalist world economy. The impression that is given from the national industrialisation projects since the eighteenth century, that has happened under the control of national governments, in national autonomy, is misleading, veiling the positions and the interdependencies of national economies in the economic world system. Presently, with continuing industrialisation, the conditions for a global transition to an environmentally sustainable society are not yet given and become more and more difficult—economic growth and growth of natural resource use, industrial production, pollution of the environment and increasing emissions of carbon dioxide in the atmosphere, with consequences of global warming and climate change, all forms of ecologically maladaptive change continue.

What happened so far in the development of modern society is not yet the transition to a post-industrial society, also not with the dramatic social changes since the end of the twentieth century—the fall of the socialist system, the breakthrough of a new wave of economic globalisation, and the technological innovations through digital information technology and the internet. What is often seen as a new society called knowledge-, network- or world society refers to partial changes in societal structures, systems and processes that do not yet sum to a new form of society, which is discussed in speculations about the social effects of the digital economy as development of a sharing economy (Bauwens 2012). These changes can be explained at sub-societal levels as changes of cultural, political and economic processes, changes of accumulation regimes, of the global political order, of international flows of people, goods and capital that have not set out of function or transformed the capitalist economic system. Industrial production, that has for a certain period dominated in the Western countries, where it was shrinking continually in the second half of the twentieth century, in terms of restructuring of the first (agriculture), second (industry) and third (service) economic sectors, has not ended with globalisation. Only the international division of labour changed through relocation of industrial production to newly industrialising countries and to the global south that are now the locations of increasing pollution, but their industrial production is mainly for export to western countries. The new industrialisation continues in similar forms as before in Europe, polluting the environment, disturbing the functions of ecosystems, increasing emissions of carbon dioxide, causing climate change and undermining the natural resource base of economic growth and development. The emission patterns at country levels are changing—since the beginning of this century the larger part of carbon dioxide emissions is from the newly industrialising countries.

From the dis-simultaneous and contradicting processes of social change in the late modern society derive controversial diagnoses that are discussed further in the following chapters as changes in global governance and economic globalisation. The picture of partial changes and partial transition processes, of new development and of modifications of earlier forms, of a society in continuous change and development, does not give a clear impression of the directionality of the processes observed: Whether these are continuing processes of economic development, or beginning transformation processes towards another economy and society. Simple distinctions of development phases such as early and late modernity, first and second or reflexive modernisation (Giddens and Beck) do not theoretically interpret the social changes further with regard to societal transformation. These distinctions of phases can be interpreted as phases of development of the industrial society, economically seen in terms of new accumulation regimes and shifting spatial patterns of industrial production that happens now primarily in the global South. The global north maintains its privileged economic position as developed and core countries, although the differences between core and periphery begin to reduce and change with globalisation. What has changed are the older constellations of actors, political and economic power relations, technological innovations and the building of a global society that develops as "partial society" at international levels, in the new forms of global communication and networking, in the development of networks of global cities and of new global classes in the international economy.

Already in the classical political economy in the early nineteenth century John Stuart Mill and others saw the end of industrialisation as caused by growth and pollution or overuse of natural resources. In the twentieth century the discussion intensified. After the paradigmatic discussions about the crisis of industrial society (Birnbaum 1969) where a blocked development or a gradual transition to a post-industrial society (Bell 1973) were diagnosed, in forms of change from an economy based on manufacturing to one based on information and services, the follow-up discussions did not create consensus. The future of the industrial society is still an open question. With the environmental movements and the "limits to growth"-debate started a broader critical discussion, including that of sustainable development with the formula for a future of industry "producing more with less". With the development of the internet the future is discussed as transformation of production from "Big Iron" to "Big Data". These are more fragments of new ecological or technological ideologies than analyses of social and economic change. What has happened instead of a transformation of industrial society since the early debates is described by Vogt (2015) as the development of the concept of post-industrial society from utopia to ideology. This seems the result of the diagnosed transitions that did not happens and require further reflection (Wildschut 2017). Empirically seen societal and economic changes happened only partially and selectively, they cannot be generalised and projected as global trends of development. At the core of many diagnoses are the phenomena of technological change, of the internet and computer-based information technology, of restructuring of economic sectors and of class structures with the rise of technical intelligence in advanced industrial societies of the west. The new economic globalisation since the 1970s builds on these developments and added new facets to it: The rise of an international technically specialised labour class, the relocation of industrial production form the old industrial to the newly industrialising countries and the emergence of a global digital economy. But the diagnoses of a technological revolution as that by Schwab (2016) from the World Economic Forum, counting four industrial revolutions in technological terms (the steam engine and mechanised production; electric energy and mass production; automated production through electronics and information technology; and the digital revolution as a fusion of technologies that blur the lines between physical, biological and information processes) are rather diffuse and iridescent as sociological descriptions of societal changes. They belong more to the wishful thinking, the visions and ideologies of a new society than describing societal transformation. The picture of the globalising society since the turn of the century is one of dis-simultaneous, partial, contradicting and conflicting changes where different trends can be identified: the continuation of older conflicts and the rise of new ones, of old and new inequalities, of technological advances and social relapses, of crises of political democracies and de-civilisation processes with new forms of barbarianism, ethnic conflicts, wars and civil wars. It seems that the inequalities, the unsolved problems, conflicts and relapses are more effective as blocking factors than as factors of a transformation of society.

3.4 Discussion and Conclusion—Knowledge Integration in the Theory of Modern Society

The history of the theory of modern society shows examples for making use of natural-scientific concepts, theories, methods and explanations—for example, in classical political economy, in the theory of Malthus, in positivist epistemology, in evolutionary sociology, in systems theory and analysis, and in the form of theorising that Horkheimer described as traditional theory in the epistemological and methodological forms of natural-scientific theorising. Yet, in the environmental research of the past decades not much knowledge integration and synthesis across the boundaries of natural and social sciences happened, only limited forms of concept transfer and knowledge exchange, or metaphorical use of natural-scientific terms in the theory of society. Natural-scientific knowledge was used in earlier theory of modern society in analogies and in concepts transfer, in Spencer's evolutionary sociology and in the theory of Marx, where, for example, the concept of metabolism or that of the "metabolic rift" was developed in the analysis of modern society.

In environmental sociology discussions about the use of natural-scientific data in empirical studies have been found since its development by Catton and Dunlap (1978), without finding consensus; this can be seen in the recent discussion of the future of environmental sociology by Lidskog et al. (2015), asking whether it is fit for a global environmental sociology that can develop beyond its national and regional differences and traditions, can be unified in themes, theories, epistemologies and methodologies to address global environmental problems or global environmental institutions. The authors connect their discussion with that of environmental sociology as maintaining a disciplinary identity or becoming interdisciplinary across the boundaries of social and natural sciences. Their answers are difficult compromises, formulated as rules for a global environmental sociology (Lidskog et al 2015: 356f):

- analysing local and place-based problems while understanding their global embeddedness and co-construction—understanding global problems as including place-based specificities, practices and effects;
- understanding nationally and regionally specific research traditions and approaches but explaining and reflecting their contextual emergence, asking whether they contribute to cross-boundary knowledge sharing and learning, which is understood as contributing to the cosmopolitan perspective Beck demands;
- the specific objects of environmental sociology, society–environment interactions, are always investigated in relation to general sociology and building on the wider discipline of sociology, arguing with Redclift that globalisation brings sociology back to the origins of social theory;

- concepts and approaches from other social-scientific disciplines are used, but collaboration with natural scientists should be practiced more cautiously, only when the social dynamics of transferred concepts are studied and the discipline of sociology is not dissolved;
- global research networks, journals, conferences, platforms, funding and audiences should not replace national traditions, but complement them;
- a critical global environmental sociology maintains the traditions of concern for environmental problems and solidarity with the victims of pollution and resource-extraction; it is a form of public sociology in the sense of Burawoy, communicating with other disciplines and non-academic institutions, without abandoning its reflective and disciplinary character.

The insufficient interdisciplinary opening of environmental sociology and the lack of discussion of the problems of social–ecological transformation of the industrial society in most sociological theories of modern society seem to go together, indicating the limits of the social-scientific debates about the future of modern society. In the broader environmental discourse, in environmental research and in the debates in environmental movements or environmental politics, natural-scientific and interdisciplinary knowledge and concepts are used more widely, especially cybernetic thinking, or biological thinking, or complexity theory (Norberg and Cumming 2008). This goes, however, together with an ignoring of large parts of social-scientific research, in attempts of understanding modern society without a theory of society or empirical knowledge from sociological research. The blocked thinking about the future of modern society, in the natural and social-scientific debates, gives rise to the questions: How can more interdisciplinary knowledge integration, for example, the integration of political-economic systems analysis of modern capitalisms and ecological systems analysis be realised; what are the epistemological and methodological requirements of interdisciplinary analyses of modern society; what are potential future development paths of society—how can the future begin?

That modern society cannot end and "the future cannot begin" was a critical reflection by Luhmann (1973), analysing the temporal structures of modern society as such of a society that blocks its transformation because of the complexity it has reached in its social organisation. In this interpretation of system complexity a transformation of modern society appears as impossible. However, the analysis remains super-ficial, reasoning with a limited set of ideas, for which it is difficult to see, where the theoretical reflection ends and the ideological thinking begins. It was not the beginning of sociological reflections about the future of modern society, but the beginning of a theoretical reconstruction of modern society that resulted in more and more reductionist and superficial descriptions and explanations of the social dynamics, sometimes masked with the concept of complexity as in Luhmann's theory. The future of the modern society is sometimes described as "the end of history", a notion known since Hegel's philosophy of history and Spencer's sociology where the evolutionary view of industrial society was that it continues forever. Such thinking was renewed after the collapse of East European socialism in views as that of Fukuyama. In Luhmann's sociological theory similar conclusions can be found, although not in a philosophy of history, in more theoretical reflections about the mechanisms of societal change, arguing that the complex global system of society cannot be changed or modified, its complexity makes all attempts to change society impossible. With such theoretical reasoning the concept of society becomes similar to that of nature, adopts a main quality of the concept of nature in earlier debates: nature as the eternal order of the material world that can change only to a limited degree and very slowly; it creates in the very long run of its existence only a perfection of its complexity.

At the end of the present discourse about modern society and its relations with nature questions come up that that were ignored in science and in the discourses about the environment. One of these questions is that of the ideological nature of science, of science as being or becoming ideological, in different forms of ideological scientism. The ideological qualities of science and scientific knowledge are not a new debate, accompanying the development of critical theory since the beginning with Hegel and Marx. Although the concept of ideology seemed outdated as analytical and critical concept with the discourses of structuralism and postmodernism and the end of the debate of political ideologies since the collapse of socialism, ideological knowledge production and application did not end, are resurfacing in other forms. The distinction between political ideologies and objective scientific knowledge showed its simplification since long, more recently in the "science wars" around the turn of the century, aiming at a critique of value-loaded social sciences, directing away from the cognitive problems of natural-scientific and environmental research. Scientific research requires the critical reflection of its knowledge practices, whether it is discussed in terms of ideology or not. With the questions of genetic modification of plants, animals and humans, of genetic engineering and cloning, ideologies, ethical and moral questions come back to science. In the critical theory discourse in the social sciences, in one or another form, these questions have been discussed; in the conventional social sciences and in the natural sciences the questions were more or less excluded and critical voices marginalised. The postmodernist discourse that developed through cultural relativism and deconstruction of social-scientific knowledge, in the debunking of "grand narratives" of modernity, has not contributed to a more critical debate, rather developed to a form of ideological scientism. Among the critical voices warning early that the development of science with the lack of ethical reflections ends in barbarized, brutal forms of knowledge technologies was Chargaff (1988).

In recent debates about the present problems of the globalised capitalism it was argued: Before the collapse of socialism and during its historical development capitalism did not need an ideology, could be defended otherwise; since the fall of socialism and Marxist political ideologies the globalising capitalism needs to defend through new—non-political—ideologies that are closely connected with the technological innovations and their economic concomitants—briefly said: the ideology of the digital revolution. The digital ideology allies with another one, the neoliberal economic ideology. The critical analysis of both ideologies, their social background, aims and effects is not yet advanced, although the discourse about modern society is now back to the question that seems to become superfluous after the fall of East European socialism and the triumphing voices of the "end of history":

How will the capitalism end? (Streeck 2016). Answers can be expected from a renewal of the discourse of critical theory, with more systematic analyses of the present development and change of the capitalist system, and from the development of interdisciplinary research about nature and society in social, human and political ecology, that bring a renewal of the critical debate of sustainability with that of socio-ecological transformation.

References

Alford, C. F. (1985). Nature and Narcissism: The Frankfurt School. *New German Critique, 36,* 174–192.

Altvater, E. (1992). *Der Preis des Wohlstands oder Umweltplünderung und neue Welt(un)ordnung.* Münster: Verlag Westfälisches Dampfboot.

Amstutz, M., & Fischer-Leskano, A. (Eds.). (2013). *Kritische Systemtheorie: Zur Evolution einer normativen Theorie.* Bielefeld: Transkript Verlag.

Baker, P. A. (1962). The Application of Ecological Theory to Anthropology. *American Anthropologist, 64*(1), 15–22.

Bauwens, M. (2012). *Blueprint for a P2P Society: The Partner State and Ethical Economy.* http://www.shareable.net/blog/a-blueprint.-for2p-institutions-the-partner-state-and-the-ethical-economy-0. Accessed 5 Oct 2017.

Bell, D. (1973). *The Coming of Post-Industrial Society: A Venture in Social Forecasting.* New York: Basic Books.

Birnbaum, N. (1969). *The Crisis of Industrial Society.* Oxford: Oxford University Press.

Bühl, W. (1983). Die 'Postindustrielle Gesellschaft': eine verfrühte Utopie? *Kölner Zeitschrift für Soziologie und Sozialpsychologie, 35*(4), 771–780.

Burkett, P. (2006). *Marxism and Ecological Economics: Toward a Red and Green Political Economy.* Leiden and Boston: Brill.

Castaneda, F. (1987). La crisis de la epistemologia. *Revista Mexicana de Sociologia, 49*(1), 13–31.

Castells, M. (2013). *Communication Power.* Oxford: Oxford University Press.

Catton, W. R., & Dunlap, R. E. (1978). Environmental Sociology: A New Paradigm. *The American Sociologist, 13,* 41–49.

Chargaff, E. (1988). *Abscheu vor der Weltgeschichte. Fragmente vom Menschen.* Stuttgart: Klett-Cotta.

Chen, X. (2017). *The Ecological Crisis and the Logic of Capital.* Boston: Brill.

Cummings, L. (2002). Why we need to avoid theorizing about rationality: A Putnamian criticism of Habermas's epistemology. *Social Epistemology, 16*(2), 117–131.

Delanty, G. (Ed.). (2006). *Handbook of Contemporary European Social Theory*. New York and London: Routledge.

Dickens, P. (1996). *Reconstructing Nature: Alienation, Emancipation, and the Division of Labour*. London and New York: Routledge.

Dickens, P., & Ormrod, J. S. (2007). *Cosmic Society: Towards a Sociology of the Universe*. London and New York: Routledge.

Foster, J. B. (2002). *Ecology Against Capitalism*. New York: Monthly Review Press.

Foster, J. B., & Burkett, P. (2016). *Marx and the Earth*. Boston: Brill.

Fuchs, C., & Mosco, V. (2016). *Marx in the Age of Digital Capitalism*. Leiden and Boston: Brill.

Gorz, A. (1982). *Farewell to the Working Class*. London: Pluto Press.

Gouldner, A. (1970). *The Coming Crisis of Western Sociology*. New York: Basic Books.

Habermas, J. (1973). *Theory and Practice*. Boston: Beacon Press.

Habermas, J. (2006). Political Communication in Media Society: Does Democracy Still Enjoy an Epistemic Dimension? The Impact of Normative Theory on Empirical Research. *Communication Theory, 16*, 411–426.

Hirsch, J. (1990). *Kapitalismus ohne Alternative*. Hamburg: VSA.

Inglehart, R. (1977). *The Silent Revolution: Changing Values and Political Styles Among Western Publics*. Princeton, NJ: Princeton University Press.

Kuhn, T. S. (1962). *The Structure of Scientific Revolutions*. Chicago: University of Chicago Press.

Lenski, G. (1988). Rethinking Macrosociological Theory. *American Sociological Review, 53*, 163–171.

Lenski, G. (2005). *Ecological-Evolutionary Theory: Principles and Applications*. Boulder: Paradigm Publishers.

Lidskog, R., Mol, A., & Osterveer, P. (2015). Towards a Global Environmental Sociology? Legacies, Trends and Future Directions. *Current Sociology, 63*(3), 339–368.

Ludwig, C. (2013). *Kritische Theorie und Kapitalismus: Die jüngere Kritische Theorie auf dem Weg zu einer Gesellschaftstheorie*. Wiesbaden: VS Verlag für Sozialwissenschaften.

Luhmann, N. (1973). The Future Cannot Begin: Temporal Structures in Modern Society. *Social Research, 43*(1), 130–152.

Luhmann, N. (1986). *Ökologische Kommunikation*. Opladen: Westdeutscher Verlag.

Luhmann, N. (1998). *Die Gesellschaft der Gesellschaft* (2 Bände). Frankfurt: Suhrkamp.

Manuel-Navarrete, D., & Buzinde, C. (2010). Socio–Ecological Agency: From "Human Exceptionalism" to Coping with "Exceptional" Global Environmental Change. In M. Redclift & G. Woodgate (Eds.), *The International Handbook of Environmental Sociology* (pp. 306–336). Cheltenham: Edward Elgar.

Maturana, H. R., & Varela, F. J. (1980). *Autopoiesis and Cognition: The Realization of the Living*. Dordrecht, Boston, London: Reidel.

Moran, E. (1990). *The Ecosystem Approach in Ecology*. Michigan: The University of Michigan Press.

Moscovici, S. (1968). *Essai sur l'histoire humaine de la nature*. Paris: Flammarion.

Norberg, J., & Cumming, G. (Eds.). (2008). *Complexity Theory for a Sustainable Future*. New York: Columbia University Press.

O'Connor, M. (Ed.). (1994). *Is Capitalism Sustainable? Political Economy and the Politics of Ecology*. New York: The Guilford Press.

Ortiz, J. C. M. (2008). Crisis y evolución actual de la epistemologia. *Coherencia, 5*(9), 169–190.

Peters, R. H. (1991). *A Critique for Ecology*. Cambridge, New York, and Melbourne: Cambridge University Press.

Rule, J. B., & Besen, Y. (2008). The Once and Future Information Society. *Theory and Society, 37*(4), 317–342.

Schwab, K. (2016). *The Fourth Industrial Revolution*. Geneva: World Economic Forum.

Streeck, W. (2016). *How Will Capitalism End? Essays on a Failing System*. Brooklyn: Verso Books.

Vogt, K. C. (2015). The Post-industrial Society: From Utopia to Ideology. *Work, Employment & Society, 30*(2), 366–376.

Watts, D. J. (2004). The "New" Science of Networks. *Annual Review of Sociology, 30*, 243–270.

Wildschut, D. (2017). The Need for Citizen Science in the Transition to a Sustainable Peer-to-Peer-Society. *Futures, 91*, 46–52.

4

Policy Change: Crisis of Environmental Policy and Global Governance

Environmental policy and governance are in a critical state, for various reasons, not only because of policy failure, weak government, and deficits of implementing environmental policy programmes, the superficial explanations found in empirical policy research. Environmental governance needs to deal with interconnected and complex problems in different social spheres, in politics, in the economy, in science, and in the social practices of natural resource use, connecting the regulation of processes in social systems with that in ecosystems. A reconstruction of the crisis of environmental governance with interdisciplinary knowledge shows other reasons than policy-related, why global agency is limited and the transition to sustainability is blocked. With the discussion of the crisis and the limits of environmental governance the significance of different forms of socialcreating knowledge for socialscientific and other knowledge for improving governance can be shown: how global environmental governance can be renewed with the help of interdisciplinary knowledge syntheses and changing knowledge practices in research and in the governance process itself.

© The Author(s) 2019
K. Bruckmeier, *Global Environmental Governance*,
https://doi.org/10.1007/978-3-319-98110-9_4

4.1   Different Views of Environmental and Political Action

In the scientific and political discourses about environmental politics and governance since the 1970s, when global environmental change was discussed, the nature of environmental governance remained controversial, more than because of old political ideologies and cleavages because of the experiences, new insights and institutional changes required. The solution of environmental problems as a trans-political process depends on political cleavages and values as reformist and transformative action, anthropo- and bio-centric values. The—naïve—view of collective learning as steady progress in building cooperation, consensus and collective action based on the policy research ignores further necessary research and the necessity of institutional changes. Controversial reasoning about the relations between environmental problems and political action unfolds in incompatible views (Table 4.1), most of them based on the solid knowledge acquired through experiences and collective learning. In the final analysis, the controversies are about the development of a collective subject of environmental governance and of global agency.

Forms of collective action are also controversially discussed in the theoretical discourse and in the practices of new social movements: whether the activities of social, including environmental, movements should be understood as political or cultural forms of action. The discussion shows the complicated interrelations between political and other forms of social action. Assuming a distinct cultural sphere and conceptualising "policy-free" forms of collective action does not help to detach collective action from politics, only from specific forms of political action, organisation and governmental institutions. The problems of political formalisation and regulation of environmental action come back continually in environmental and all kinds of collective action. With the diffuse concept of governance the dilemma is only partially solved, through the broadening of governmental action and public policy processes, opening them for participation of further, nongovernmental actors and forms of civil society action, enlarging the scope

Table 4.1 Differing views of the environment and politics

1. *Environmental problems are or become political* as they require power and collective and political action to be solved—the human relations with nature, the personal and private life, and other social spheres and activities that were hitherto not part of public policies and politics have now become political

2. *Environmental problems* are not primarily solved through politics or governance, but *changes of individual and collective consumption behaviour and lifestyles* that do not require politics; environmental problems are solved through cultural and behaviour changes

3. *More than in consciousness, worldviews and subjective attitudes the environmental problems are rooted in the systemic constitution of modern society*, in its capitalist mode of production and its socio-metabolic regimes that develop and change through collective action and regulation human relations with nature beyond policies and political governance

4. *Environmental action is not political action*, follows individual and cultural motives, orientations and practices of social movements and networks; social changes happen through collective and social learning

5. *Environmental action is guided by culturally specific worldviews, values and ethical principles*; social change happens through changes of worldviews, beliefs, ethical thinking and reasoning, less through scientific knowledge and political action

6. *Environmental action is socially structured, organised, regulated, institutionalized action* in which the political appears as the final form of regulation and control that directs and limits all forms of social action

7. *Global environmental governance is a false abstraction and conception of agency*—all environmental problems need to be dealt with locally, by specific actors, in concrete forms of action, in social networks, not through political centralisation and "top-down" approaches

8. *Global environmental governance is an unrealistic idea under the given conditions of a political world order* based on the national states that make international policies and regimes to weak institutions

9. *Global environmental governance is a broadening of the political space and forms of political action* through which other than governmental and political actors are included in political processes; it requires new forms of democratic legitimation and representation

Sources own compilation; Jakobsen (1999), Meyer (2001)

of governmental action, creating new forms of political action where governments and party politics reach their social and institutional limits, and to creating further legitimation for collective decisions. But with that, the boundaries between different forms of social action are blurred and it becomes unclear where the boundaries of governance are and how they can be shifted. Global environmental governance is confronted with the political forms of action and institutions that developed in the epoch of modernity and their limits, the modern nation state that developed in political modernity in the form of the democratic nation state. International relations, policies and regulations, laws, conventions and agreements are created and controlled by the governments of nation states. No other legitimate political institutions exist in the international sphere, except the ones created by the states to maintain their common interests, especially that of regulating global exchange, trade and mobility. The early modern concept of nation states and national societies is outdating in late modernity, with the rapid forms of global social and environmental change experienced in the twentieth century. New international institutions develop slowly. The nascent forms of international politics and collective subjects described with the governance concept show forms of "extended politics" through co-optation, cooperation and participation of political and nonpolitical actors and organisations. Hannigan (2006) or Gross and Heinrich (2010) argue for the further development of global environmental governance in such forms. Beck discussed the concepts of "sub-politics" and "cosmopolitanism", but did not find convincing arguments for his terminology for conceptualising new forms of political action developing with environmental movements and in environmental politics.

The controversies about global environmental governance indicate a lack of consensus in science and politics about political action, about the limits of governmental action and of international institutions: a legitimation and knowledge crisis for with multiple reasons: lack of scientific knowledge, lack of political consensus, asymmetrical power relations in the international processes, interference of non-legitimised economic forms of power and political power, vested interests of global players that do not support a strengthening of international institutions. The more critical debate of governance is about the possibilities of building a network of global institutions that can act independently from the control of states and create a new space of political action where global agency and transformation capacity can develop. The inherent tensions in global environmental action are articulated in the debate of solving environmental problems,

- either through collective action reflecting particularities of culture, place and lived experience of people (Jasanoff in: Jasanoff and Martello 2004: 49),
- or through multi-scale governance in a more limited sense, developing from the earlier discussions about decentralisation and federalism in politics (Piattoni 2009).

The organisation and development of global political action and its institutions in a continuous debate is part of the governance process: environmental governance has to deal with the inability in science and politics to create long-term, intergenerational perspectives and rationalities of collective action that support the transformation of the ecologically maladaptive and dysfunctional global economy and the weaknesses of global institutions to create transformation capacity for sustainability which requires further changes of political and economic institutions. Relevant changes in the international policy arena during the past decades include

- a broadening of governmental action,
- a differentiation of the national state into local, regional, national and international spheres,
- involvement of new groups of actors, as articulated in the governance concept (connecting governmental and nongovernmental forms of action),
- and new forums and platforms of transdisciplinary knowledge production, collective learning and action.

Regarding the social problems that affect and limit global environmental governance and transformation to sustainability, the social, political and economic inequalities and divides, including the global divide between North and South in the modern economic world system, require more efforts in knowledge generation and transfer, collective and political action, generation of transformative capacity and global agency. Global environmental governance is part of a larger process of creating a global political system that develops not only new cultures of cooperation, of joint action of governments and of further actors—at this level the debate is stuck in the rhetoric of "our common future" and the future of humankind. The further capacity of regulating society–nature relations and creating global agency begins with creating social innovations and institutional change, in the discourses about new normative orders, political legitimation, and democratic participation, in discussions of political and social cosmopolitanism, about environmental justice and environmental citizenship. What the developing global environmental governance discourse can do to support the creation of a new normative and political order in the international arena can be described as the normative framing of global governance; it requires the creation of a new global "political society" (called society here because it is not a state) and of a global civil society (Jasanoff in: Jasanoff and Martello 2004: 31ff).

4.2   Global Environmental Governance and Earth System Governance

The discourse of global environmental governance developed in connection with the discourse about sustainable development and was restructured with the concept of Earth system governance (Biermann 2007; Biermann and Gupta 2011). The earlier debate about global environmental governance was summarised in a publication from the International Institute for Sustainable Development in Winnipeg (Najam et al. 2006; a more recent overview is given in Pattberg and Zelli 2015). The discourse is institutionally centred in a long-term international research programme, the Earth system governance project, located at the University of Lund in Sweden.

1. *Global environmental governance* (Pattberg and Zelli 2015) and global environmental politics (Dauvergne 2012) are not single approaches or paradigms; global environmental governance encompasses a plurality of ideas, approaches, perspectives and institutions. In discussing these, the common ideas, the aims, premises and practices, the deficits and limits of global governance, become visible. The global

governance project emerged at time and in a crisis situation in environmental research, a latent crisis of the environmental sciences that was not articulated openly, only indicated as a knowledge-related crisis in some events and aspects: the debate about post-normal science (Funtowicz and Ravetz 1993), about new knowledge production (Nowotny et al. 2001), about the interdisciplinary opening of the social sciences social sciences (Gulbenkian Commission on the Restructuring of the Social Sciences 1996). As parallel development can be seen the rise of the resilience discourse connected with the movement of the resilience alliance (Folke 2006), that implied changes in the sustainability discourse through which the guiding idea of sustainable development itself is modified and becomes controversial.

Global environmental governance is discussed in other forms as Earth system governance by the authors in Pattberg and Zelli (2015), in debates of the reform and institutional development of the international system, in policy-close forms. The aim is to create knowledge for dealing with environmental change in the perspective of sustainable development.

2. *Earth system governance* was described by Biermann (2007) as a new paradigm that developed in the context of sustainability science (Table 4.2).

The ideas of global environmental and earth system governance were repeatedly criticised because of the dominance of Western scientists from OECD-countries, and because of the top-down approaches that rely, unrealistically, on the international political institutions and international regimes to manage and control the earth system and the climate system. This critique can be seen as reflecting the inadequate international institutions at the beginning of the global governance process and the necessity to learn ways out of the dilemmas, traps, limits of the old global order.

Pelayo (2008) described the crisis of global environmental governance as one caused by accelerated and probably irreversible human modification of nature, as a trans-sectorial crisis that cannot be limited to a policy crisis and requires collective action by the whole

Table 4.2 Global governance and earth system governance

1. *Global governance—new forms of international politics*:
  • new, nongovernmental actors (e.g. environmental movements) and forms of civil society action
  • new organisational mechanisms (international regimes, private–public partnerships, market-based policy mechanisms)
  • multilevel governance (connecting local, national, global levels of cooperation and governance)
2. *Earth system governance—a paradigmatic project of the twenty first century*:
  • protecting the entire system earth and its subsystems (social and ecological systems)
  • building stable institutions to enable a safe transition and co-evolution of natural and social systems at planetary scale
  • broadening the kinds of actors participating in governance: governments, civil society actors, local actors, national alliances, public and private actors
  • creating institutions for effective international collaboration

Global environmental governance, a less standardised term, shows similarities with earth system governance with regard to the underlying assumptions and views of the nature of environmental governance and of policy perspectives. Whereas earth system governance is more clearly oriented towards the social–ecological transformation of coupled social–ecological systems, both debates about environmental governance and its development require further, more critical debates and knowledge about the social, cultural and economic conditions, processes and systems that form their contexts. The inherent inclination to accept the influences of neoliberal globalisation on all governance processes as a reality taken for granted becomes the critical point in further discussions

Sources Biermann (2007), Biermann and Gupta (2011), Biermann and Pattberg (2012), Biermann (2014), Pattberg and Zelli (2015), Kanie and Biermann (2017)

human community. The argument that the whole human community needs to act is convincing in a theoretical reasoning, but the real difficulties remain, how the human community can act collectively; this continually controversial question cannot be answered by appealing to "our common future". Pelayo describes also the long-lasting or irreversible effects of global environmental change, climate change, species reduction, ocean and air pollution, as requiring answers in science and politics. But the discussions did not advance to strategies for building global agency.

4.3 The Crisis of Environmental Governance in Theoretical and Empirical Perspectives

The crisis of global governance can be read off from the difficulties experienced in global environmental policies like climate policy and from the controversial discussion about the institutions of governance. The policy crisis indicates a deeper crisis in which social, political, economic, scientific and environmental problems and conflicts combine and overlap; it can more adequately be called a global crisis of the regulation of social ecological systems, a multifaceted knowledge, system and governance crisis, as a "polycrisis" (Morin and Kern 1999: 14; Benington and Moore 2011) to be analysed in the following aspects (Table 4.3).

The different facets of an ecological system crisis have been discussed since longer time, as different and specific phenomena of crises related to changes of the capitalist economy that appeared in connection with the globalisation process: the limits of fossil energy sources, the changes of the global energy system that is based on the limited fossil energy resources and the forms of global environmental change.

Table 4.3 Crisis of environmental governance—connections with further crises

1. *Problems of policy failure*: the necessity and the form of a world environmental organisation (Biermann 2007)— world party, the development of new political subjects and actors, new forms of a global commonwealth
2. *Problems of market failure*: deregulated markets and their allocative and distributional effects
3. *Development of a global civil society and the crisis of democracy* in which problems of inequality resurface: "digital divide"
4. *Crisis of science or knowledge crisis* (Jennex 2014: 2ff) in dealing with the complexity and the limits of knowledge, from which resulted the discussion of new forms of knowledge production (Nowotny et al. 2001), in the social sciences and their future (Lee 2000)
5. *Deficits of knowledge integration from the social and natural sciences in environmental* research and competing theories and approaches of knowledge use in environmental governance (Pahl-Wostl et al. 2013)
6. *Systemic nature of the crisis*: discussed by Wallersteins (2000) with the guiding question about the long-term trajectory of the world system—globalisation or transition?

Sources own compilation; sources mentioned

In the discourse of critical theory of society several efforts have been made to describe the system crisis more systematically. Marshall and Goldstein (2006) developed a conceptual crisis model for late capitalist societies to systematise the analysis of different kinds of crises: accumulation crises of over- and underproduction, a social welfare legitimation crisis, and an environmental legitimation crisis. The crises are reconstructed with a simplified model of system components and functions of interacting systems in modern society (economic system, political-administrative system, sociocultural system and ecological system). The functional analysis does not reveal the internal structures and inconsistencies of the capitalist system and its mode of production and reproduction. Therefore, the model is limited for theoretical and explanatory purposes; it draws on authors from the Frankfurt School, especially Habermas, trying to explain legitimation crises with regard to social welfare and solving of environmental problems. The interactions between social and ecological systems, society and nature, require further analyses; superficial diagnoses as the failure of the political-administrative system or the government ignore the nexus of social and ecological systems as a crisis-generating mechanism.

More systematic analyses of the capitalist system crisis are found in the discourse of ecological Marxism (O'Connor, Altvater, Foster, Burkett). The analyses show the complicated systemic mechanisms that evoke interconnected crises in late capitalism. This critical discussion happens in a closed theoretical discourse of Marxism, hardly taken up in inter-theoretical and interdisciplinary knowledge syntheses and remaining limited in influencing in the ecological discourse. A variant of critical theory more open towards inter-theoretical and interdisciplinary knowledge integration is world system theory. In this discourse a theoretical reconstruction of system crises and a diagnosis of linked crises in the modern world system has been formulated by Moore (2011) in the theoretical analysis of the capitalist world ecology, and by Chase Dunn (2013) who distinguishes five interconnected crises of hegemony and global governance, inequality and democracy, human–nature relationship, the capitalist economic system and a political crisis of the new global left. With this diagnosis, it is possible to understand the limited success of global governance as one that cannot be reduced to single factors as policy failures described in empirical policy research, in the last analysis, as caused by the specific nature of the modern economic world system.

The discourse of political ecology, drawing on similar knowledge about modern capitalism as ecological Marxism and world system theory, is more diffuse and theoretically inexact; it includes several variants, connected to different disciplinary knowledge. The development of political ecology, especially in the variants connected to geography (Walker 2005: 79), shows the theoretical difficulties to connect and integrate knowledge from different theories and disciplines in environmental research. It remains unclear in Walkers review, how knowledge can be integrated; he refers to the difficulties of the new subject through its changing aims and cognitive interests, incoherence and sprawl, as a patchwork of vaguely connected ideas and approaches, where the question, which knowledge is used for which purposes in the interdisciplinary synthesis, is answered differently. Such forms of disjointed incrementalism (documented in its variations in Bryant 2015), consequences of varying empirical knowledge bases and influences of different cognitive and political interests, reduce the analytical capacity of the approaches that appear finally as variants of political activism, or, in the description of Forsyth, as the politics of environmental science (Forsyth 2003).

The discourses of ecological Marxism, world systems theory, world ecology, and political ecology are not reconstructed here in detail; only parts of their concepts and theoretical analyses are used in the construction of a social–ecological theory of nature and society that serves as integrating theory for the analyses of the limits of global environmental governance (Chapter 7). In the following discussion, the crisis and the limits of global agency and environmental governance are reconstructed in the different perspectives and analyses summarised in Table 4.3.

1. Policy failure is not the failure of single policy institutions that can be empirically studied in terms of effectiveness or efficiency, related to bureaucratic or other inefficient organisational structures, implementation deficits etc. The failure is a consequence of the systemic constitution of modern society with systems like the state and the capitalist

economy, and the division of functions and labour between these systems. Improvements cannot consequently be done through corrections of single institutions or institutional development, requires transformations of the systems which again requires critical analyses of their dysfunctions and maladaptation. Without further knowledge than that created from exemplary empirical research in case studies or policy evaluation, it is impossible to improve global environmental governance.

2. Market failure refers to the systemic constitution of the modern economic world system where markets have a main role. The dysfunctionality of markets with regard to distribution and redistribution of natural and other resources according to the needs of all is generated through the monetary mechanism that mediates market-based exchange. The basic human needs are defined for practical purposes, for example, in the "human development"-index of the United Nations, and needs as food, clothing, housing, care, education require more and other distributive systems than markets, also other forms than an economy of planning that has lost persuasiveness since the collapse of east European socialism. The embedding of markets in earlier societies or other economic institutions of a redistributive economy (reciprocal exchange and gifts) compensated market inefficiency. In the modern economic systems, such compensating and controlling mechanisms are missing; the deregulation of markets through globalisation is even increasing the economic maladaptation and ecological maladaptation of the global economic system.

3. The difficulties of building a global civil society articulate the democratic deficits in the processes of governmental policies, in economic and political globalisation. The informal processes in which participation and stakeholder involvement happen in global environmental governance and in global environment assessment cannot count as good examples for civil society action and for developing new cultures of democratic decision-making and politics in the international arena. Too much in these informal processes of stakeholder co-optation and involvement is unclear. The development of the global civil society depends on the development of civil society in the countries, and the forms of action and actors that create the local civil societies. To

bring the discussion away from the limited forms and experiences with stakeholder participation, it would be necessary to connect the development of a global civil society with other ideas, institutions and processes: the building of new normative orders for framing international and global governance, and the development and formalisation of participation rights larger social groups.

4. A crisis of scientific knowledge production towards the end of the twentieth century began with the epistemological discourse about new knowledge production, "mode 2" or "transdisciplinarity", which centred on the organisation of research and its specialisation, coinciding with a crisis in the theory of society in sociology (see Chapter 2). The epistemological problems of knowledge production, problems of interdisciplinary knowledge generation, attempts to theorise society with the cultural relativism in the "small narratives" of postmodernism, and efforts to develop a new theory of the globalising economy and society disturbed the routines of academic and disciplinary research (Gulbenkian Commission on the Restructuring of the Social Sciences 1996). As the other facets of the global crisis, the crisis of science deals with the difficulties that can in abstract form be formulated as clash of different knowledge cultures, such as particularism and universalism, instrumental and value rationality, or Western and Eastern science and thinking. The last two "grand theories" of modern society, the system theory of Luhmann and the critical theory of communicative action of Habermas, were shortly after their completion criticised as exemplars of an outdating form of theory of modernity that dealt more with the rearranging of theoretical thinking in sociology than with opening the theoretical discourse for the analysis of the coming conflicts and problems in late modern society. Since then sociological theory of society is becoming less influential in many debates, also in the environmental discourse.

In the discourse of postmodernism, the idea of "grand narratives" of modernity was deconstructed through epistemological reasoning, resulting in many differing small narratives. The handbook of European social theories (Delanty 2006) marks the rupture in theory construction through that came with the diffuse notion of social theory where the analysis of modern society is fragmented in a number of partial, thematically specialised theories. Attempts to renew the discourse of a critical theory of modern capitalism in sociology failed so far. Swedborg (2014) criticised the bad state of theoretical reflection in the social sciences more generally, however seeking for a renewal in going back to older theories and authors, without renewing and broadening the theory of modern society. The internet-based digital ideology of "Big Data" developed with the message that theory has become superfluous with the new knowledge technologies that can produce empirical data in the magnitude "n = all". Ecological research went through crisis diagnoses as imperfect science (Peters 1991) and several efforts of renewal and interdisciplinary broadening in the past decades without advancing towards a new synthesis in a theory of society and nature. Suggestions about integrating social- and natural-scientific environmental research (Strang 2009) remain preliminary, incomplete and not practically applicable for a transformation of society and economy, and so remain discussions of an "economy of the earth" (Sagoff 2008), and the policy-related efforts to deal with the environmental legitimation crisis (Marshall and Goldstein 2006). The efforts to global knowledge synthesis about the state of the environment (Millennium Ecosystem Assessment, the reports of climate research by the IPCC) did not yet result in improvements of global governance, only providing improved knowledge about the worsening environmental situation and the deficits of environmental governance.

The connected crises were mainly discussed and articulated in the social sciences, also the environmental aspects, whereas the natural sciences are much less involved in the diagnoses and analyses of governance problems, which seems a consequence of their conventional, discipline-specific forms of research and theoretical reflection. This disciplinary fragmentation and inequality in knowledge application needs to be analysed further in connection with the review of the epistemological discourses of inter- and transdisciplinarity.

5. How far is the integration of the social and natural sciences in environmental research and production of knowledge about global change and for environmental governance reflected? Pahl-Wostl et al. (2013) diagnosed, in spite of the longer debates about inter- and

transdisciplinary approaches, only slow progress in data and knowledge integration and give as reason methodological problems: lack of innovative methodologies and new methods and results in the conclusion to develop new forms of knowledge bridging and integration, making scientific knowledge more policy-relevant. The methodological problems are discussed further (see Chapters 5 and 6), reflecting more the epistemological questions and the relevance of theoretical knowledge for further knowledge integration. The presently intensifying discussion of social–ecological transformation channels these methodological and epistemological debates in the perspective of a transition to sustainability.

6. The systemic nature of the crisis is articulated with the question of Wallerstein whether the present forms of global change should be understood as globalisation or as part of the long-term transformation of the modern world system. This question reveals the great divide in the controversy about the present crisis and its solutions. For Wallerstein, the globalisation discourse implies a misreading of the historical reality of the modern capitalist world system, that is unfolding as an incoherent and conflicting global system since the beginning of the European colonialism in the sixteenth century and the concomitant division of the world in different parts with unequal development. Unlike other descriptions of the long societal modernisation process, the division of the world in two separate forms of developed and non-developing economies, in countries with advanced and delayed development, is refused in world system theory, where both are seen as necessary parts of the same system of modern capitalism.

The modern world system is a system that generates and reproduces this global divide between North and South as a necessary component of its functioning and development. Throughout its history modern capitalism developed as an imperfect system in inequalities, and conflicts—the territorial components of economic centre and periphery, the social and economic inequalities, the class conflicts between labour and capital, and conflicts between different capital fractions in the accumulation process that are temporarily stabilised in forms of accumulation regimes driving the economic development in "long waves". Although the present globalisation process integrates with the industrialisation further countries in the growth mechanism, none of the problems with social and economic inequality and the rapid deterioration of the environmental quality is solved or beginning to be solved. Wallerstein, who refuses to use the term theory for world system analysis, that it requires essential components from macrosocial, critical theories of the capitalist political economy and of the modern society. In difference to other views of the globalisation process, world system analysis characterises the modern world system as an economic system, whereas the development of a world society, according to other theories already existing, is seen as more complicated. The world system is a partial society developing in heterogeneous forms, with two contrasting components of centre and periphery, with the political institutions of nation states that do not develop towards a global state or government, and with manifold cultural differences that do not vanish. From this diagnosis begins the further discussion.

The differentiation between policy crisis and systemic crisis provides two contrasting views of the present crisis—either one of policy and market failure that needs to be cured through policy reforms, or one of a transformation towards sustainability that implies necessary regime conflicts in the long process of the transition to a new society and, mode of production and socio-metabolic regime. The transformation of social–ecological systems towards sustainability is the building of global agency. This can be described in more detail by analysing the knowledge problems in global agency. The problems of global environmental governance discussed can be reformulated in two contrasting diagnoses, on of institutional failure to cope with in policy reforms, and one of transition conflicts generated through the antinomies and contradiction of the economic world system.

a. *Diagnosis of institutional failure/regulatory inefficiency*: State failure, market failure, community failure in natural resource management result in the present governance crisis, requiring new attempts to formulate principles of good governance learning from past failures. *Dominant views* of the problems in environmental governance argue in this sense, unfolding in a chain of arguments:

- existing institutional arrangements in natural resource management have failed (state-based approaches, market-based approaches and local community-based approaches) and need to be reformed and improved;
- the knowledge for such improvements is created in normal science, through specialised and empirical research, with a gradual improvement of institutional arrangements;
- the reform process can be conceptualised theoretically as ecological modernisation of global governance: political reforms and coordination, institutionalisation of ecological rationality and complicated networking of institutions as strategy to achieve sustainability; a green economy is achieved in this way through institutional corrections, within the capitalist economy;
- nobody knows a master strategy for transition; transition happens through collective learning (to cooperate, to change worldviews and interests) and change through adaptive management and governance (muddling through).

The knowledge used in such an adaptive strategy of ecological modernisation is produced in empirical policy research on the effectiveness of international environmental regimes. The results are summarised in Table 4.4.

Neither from the theoretical, nor from the empirical research on political institutions and policy processes develops a coherent strategy, more partial adaptations that resemble "piecemeal engineering" and "muddling through". Some basic principles are formulated, suggesting more or less radical reforms, institutional reorganisation and adaptation, regarding the economic internalisation of negative external effects, and changes of lifestyles, but all with diagnoses that do not see the necessity of system transformation. The theoretical backing of the policy reform perspective is limited, mainly formulated in specialised research as policy research on international relations and international regimes (see Table 4.4), and sociological research on ecological modernisation (Mol).

Table 4.4   Effectiveness of international environmental regimes

The effectiveness of international environmental regimes and agreements in terms
of solving or mitigating the problems they address is influenced by a multitude
of regime-internal and external, contextual factors. Young (2011) showed from a
comparison of hundreds of case studies the following results:

1. *General findings*:
   - The anarchic character of the international society is not always an obstacle for
     regime effectiveness
   - Regime design is more often a factor of success than the estimation of the prob-
     lems as easy/difficult to solve
   - Significant parts of regime success can be attributed to activities that are not
     regulatory in the sense of prescriptive regulation (through prohibitions and per-
     missions) but through procedural and other, including knowledge-generating
     functions
   - Environmental regimes are dynamic, changing and adapting continually after
     their formulation
   - Success of environmental regimes is highly sensitive to contextual factors
2. *Specific findings*:
   - Active participation of a single actor (hegemon) is not a necessary condition of
     regime success
   - Success in the implementation of international regimes is likely to require a
     maximum (not a minimum) coalition of winners
   - Creation and maintenance of procedural fairness and legitimacy is important to
     effectiveness and for maintaining active participation of actors over time
   - Legally binding conventions or treaties do not necessarily ensure higher levels
     of compliance on the part of the actors
   - For some environmental problems private governance and hybrid governance
     systems with public and private actors can be successful
   - For many environmental problems, multiple pathways of problem-solving can
     be successful
3. *Findings about institutional interplay*:
   - Institutional interplay is just as likely to produce positive or synergetic effects as
     it is to lead to interference between or among regimes
   - Conflicts between regimes can be resolved through negotiations, not necessar-
     ily requiring subordination of one regime to the other
   - Regime complexes offer new ways of solution where problems cannot be solved
     through the creation of a single regime
4. Conclusions
   - *Deep structure*: environmental regimes need to reflect the deep structure of
     the international society: restrictions on the sovereignty of member states and
     severe sanctions are not effective in the international arena
   - *Problem structure*: to differentiate between problems that are easier or harder
     to solve (on a benign-malign spectrum) does not always help to predict regime
     success

Table 4.4 (continued)

• *Power* as a determinant of regime effectiveness is a complex and contested issue, the conceptualisation and measuring of different forms of power is a main practical problem
• *Participation* is usually understood as following the law of the least ambitious program; this is not necessary the case in international regimes
• *Compliance*: rule enforcement through government or law is not necessarily creating compliance of the actors
• *Fairness and legitimacy*: regarding this point, there are no clear trends and generalizable results
• *Policy instruments*: no generalizable results—the knowledge required is about the conditions under which specific policy instruments prove effective
• *Interplay management*: no generalizable results for international regimes—the knowledge required is about synergy through institutional interaction and regime complexes
• *Nonlinear changes* in ecosystems are frequent, but how to deal with them in governance systems is difficult; effectiveness and adaptability need to be balanced
• *Scale:* little is known about generalizability across levels of social organisation (from local to global)
The findings and conclusions formulated by Young show similarities with earlier results and comparison of environmental and resource management regimes, for example, in the research by Becker and Ostrom (1995), Agrawal (2003), Acheson (2006), Lemos and Agrawal (2006): it is easily found out that combinations of different, policy-, market-, and community-based forms of governance may be necessary, but under which conditions, in which forms and relations, is an open question. Ex post evaluation of regimes and institutions shows case-, scale- and situation-specific success and failure.
The limits of generalizability seem to be inbuilt in the empirical research on single regimes and their comparison. Furthermore, the research about effectiveness seems to make little use of theoretical knowledge, systems analyses, qualitative and functional differences between social and ecosystems, and the conditions and contexts under which international regimes and governance systems operate (although contextual factors are among the general findings reported above, but showing only the casualty and contingency of contextual influences found in empirical research). Young's conclusions say more than the findings about the knowledge from regime research: the knowledge dilemma is that no generally valid predictions about regime and policy failure or success can be made from the accumulated data; they show more the specificity and conditionality of the perseverance of institutions and regimes. Starting from such conclusions it seems necessary to extend the knowledge creation about international and global governance through other forms than specialised empirical research: through interdisciplinary knowledge syntheses, syntheses of theoretical knowledge about society and nature.

Sources Young (2011), Becker and Ostrom (1995), Agrawal (2003), Acheson (2006), Lemos and Agrawal (2006)

b. *Diagnosis of transition conflicts and system transformation*: different types of international, national and local resource management regimes clash in the social–ecological transformation processes on the way to sustainability. The conflicts range from local resource use conflicts to conflicts about the regulation of the global economy and conflicts between international regimes. Such a critical diagnosis, based on analyses in political economy, social and political ecology, unfolds in another chain of arguments to deal with the governance crisis:

- policy reforms and improvements are necessary but not sufficient, they do not touch the core problems, rather shifting necessary changes in social–ecological systems to the future when the problems increase and solutions are more difficult;
- the social–ecological transformation towards sustainability is a conflicting process in which different resource management regimes clash (the economic and ecological regimes at international, national and local levels);
- a master strategy for the long social–ecological transformation to sustainability does not exist, is unrealistic for a multidimensional (political, social, cultural, economic) process that lasts for many generations; however, transformative strategies can be formulated that develop and change in the course of transformation, based on the research and interdisciplinary knowledge syntheses, providing knowledge for different phases of transformation; the strategies need to be reviewed and improved continually, with the advances of knowledge production, through experiences and collective learning;
- social–ecological transformation needs to start from the existing institutional regimes and complexes, go through conflicting processes in which economic and ecological resource use regimes, economic and ecological interests of different resource user groups clash, economic and ecological distribution conflicts need to be solved;
- the solution of economic and ecological distribution conflicts implies to combat and transform the neoliberal economic world

order, to transform the globalised capitalism, and through that to build new social–ecological regimes as the basis for a new sustainable society;

- the conflict solution opens ways towards transformation which is a multigenerational, multi-scale and multiphase process in many forms that cannot be foreseen at the beginning—the future sustainable society is unknown, is approached through the successive building of new social–ecological regimes.

Also, the critical views are not completely coherent; they are based on different theories, but consensus is found about a societal transformation and some principles for renewing global environmental governance. In the past decades, developed in interdisciplinary research on society and nature several approaches in human, social, political ecology and political economy, two of which develop with elaborated theoretical perspectives: world systems theory and the new social ecology; in both approaches develop the perspective of a world ecology that is based on theoretical analyses of the interaction between societal and ecological systems.

The controversies between conventional approaches of ecological modernisation or policy reforms, and critical approaches of social–ecological transformation found in national debates about environmental governance influence global or earth system governance in continuous debates about the knowledge use and practices fights on influence, definition power and social knowledge practices (see Chapters 5 and 6). The building of new social–ecological regimes requires global governance, which can be explained with ecological and social knowledge. For the theoretical development of global governance social–ecological transformation can be understood methodologically as similar to ideal types in the sociology of Weber, constructed in changing strategies of controlled change for different phases of the long transformation process. In the practice of the ecological discourses in science and policy and in the sustainability process the ideas to create transformative capacity will always need to be discussed and defended in controversies, in conflicts and hegemonic power fights. This implies to use a variety of ideas, concepts, methods, forms of knowledge bridging and knowledge sharing, in which also contradicting rationalities and strategies need to be connected. Ideas derived from natural-scientific environmental research like adaptive management and governance are relevant at the beginning of transformation, and from social–scientific environmental research certain approaches from critical policy research and system analyses of political and economic systems. Practically, seen the improvement of global environmental governance with new knowledge is to connect the limited policy reform debates with broader approaches, through additional forms of knowledge use to deal with the difficulties and unsolved environmental problems in global governance. This can be formulated in terms of conditions of success of global governance where the deficits, the illusionary visions of certain approaches, and the more critical and realistic views of the problems in other approaches are confronted with each other. Epistemologically and theoretically seen the differences of the approaches can be described as follows:

- naïve, illusionary, utopian visions and approaches are less based on system analyses of society–nature interaction, more on worldviews and normative ideas; they work with combinations of normative criteria and knowledge from empirical research and policy analyses, without theoretical backing through interdisciplinary knowledge synthesis; naïve visions of good governance and the good society are examples for such approaches, arguing with individual behaviour changes or cultural changes, or using knowledge selectively to argue for soft approaches such as reforms in policy processes, in natural resource management, and for changes of life- and consumption styles;
- more critical, scientifically backed, and complex approaches argue with the help of system analyses of the globalised capitalism and of the global interaction of society and nature in modern society; they seek for the solution of economic and ecological distribution conflicts, develop critical normative views derived from the assessment of the systemic world order, and specify the requirements of long-term transformation processes, implying transformations of societal systems.

Two critical questions for the further development and building of agendas for global environmental governance refer to the transformation of the economic word system and the interaction between modern society and nature:

1. How far are the programmes and strategies connected with such for a transformation of the capitalist world system? This economic system is dominating large parts of society and social life with its control of labour and production, natural resources, consumption and capital flows and economic accumulation and reproduction and cannot remain unchanged. This requires consequentially a strategy of the societal control of the economy, re-embedding and regulation of markets, redistribution of natural and social resources, and democratisation of economic power and processes—all missing in present global governance agendas.
2. How is the interaction between society and nature constructed and mediated through global environmental governance? This governance is the core process of societal transformation towards sustainability, a process that connects to heterogeneous and differentiated processes beyond policies to be influenced and regulated in governance processes: the transformation of the global economic system just mentioned, the changes of life- and consumption styles, the transformation of demographic reproduction processes which are in manifold ways connected with social processes in the lifeworld, in the economy, in education and culture.

Answering these questions it becomes impossible to reduce global environmental governance to the political processes of goal formulation, agenda building, programming, reform, coordination and cooperation. In more elaborate forms the complicated issue of "political nature", how nature and the political orders interact, is discussed by Meyer (2001) in a critique of environmentalism that he describes as derivative, better known as "following natures lead" or "embeddedness of human society in nature", confronting it with a dualist view of nature and polity as two distinct yet connected spheres: he developed the ideas from a re-reading of classical philosophical authors, Aristotle and Hobbes. Meyers critique of environmentalism is, that it holds neither empirically nor theoretically in attempts to build an environmentally sustainable social order. Without giving a detailed account of how to reformulate and restructure environmental governance, his reasoning is convincing in showing the deficits of naïve and utopian environmentalism that works with simple suggestions of changes, for example, the technological approach of "small is beautiful" in early environmentalism, or the normative ideas about mental changes in deep ecology. However, his arguments do not dissolve the critique of shortcut connections of normative and factual ideas in environmental action that was already found in an earlier study by Ophuls; he analysed the political, social and economic implications of the environmental crisis that was underscored not only by the majority of people, politicians and scientists, sometimes also by critical environmentalists. For Ophuls, the weak approaches reduce solutions to a limited set of radical political reforms, but remain vague at the levels of more fundamental changes and transformations of the ways of life and institutional arrangements in Western industrialised societies. His conclusion is, that ecological scarcity is the overarching problem which reformist policies of ecological management cannot solve; they can at best postpone "the inevitable for a few decades at the probable cost of increasing the severity of the eventual day of reckoning" (Ophuls 1977: 3).

The perspective in Ophuls' analysis is one of Malthusian thinking in terms of scarcity and addressing and addressing environmental solutions mainly in terms of politics, diagnosing a crisis of liberal democracy, and arguing for a new political philosophy and new political institutions. Thus, his analysis is caught in a short circle of normative reasoning, not yet connected to a systems analysis of modern capitalism in political-economic terms, although he goes some way in analysing modern socioeconomic systems, centring around the concept of a steady state economy found in classical political economy (Mill) and renewed in ecological economics (Daly 1973, with the participation of Ophuls and most of the Neo-Malthusian ecological authors around the "limits to growth"-debate). Ophuls' analysis is stuck halfway between conventional and critical approaches to environmental governance. He transformed the economic ideas into a political theory of the steady state that does not sufficiently analyse the capitalist economic system which he identifies as the blocking institutional order, using the short-cut formula of "mutual coercion, mutually agreed upon" by Hardin (Ophuls 1977: 155f). Meyer does not follow the environmentalist thinking of Neo-Malthusianism and scarcity, but does not find a way out of the dilemmas with the shortcut connections between politics and nature that remain at the level of worldviews and normative visions. The consequent step to analyse theoretically the interaction of society and nature in modern capitalist society is not done by Ophuls and not by Meyer. But both their analyses help to identify the deficits of political and environmental thinking. In abbreviated form the problem can be described as confrontation and controversy between approaches

- of simple normativity, where selected scientific knowledge is directly used to support normative ideas and normatively guided political action that are seen at the core of collective environmental action and governance, and
- critical normativity, where the normative ideas or paradigms guiding scientific analysis are connected with and integrated in the theoretical analyses of interacting social and ecological systems.

In both approaches, the necessity of normative principles and ideas in scientific research is not denied, but they arrive at contrasting views of environmental governance problems. The differences are not to describe only as political differences in terms of reformist or radical strategies or as differences of worldviews which end in the construction of cultural paradigms for environmentalism such as the dominant western worldview, the human exceptionalism paradigm and the new ecological paradigm in early environmental sociology by Catton and Dunlap. The political sphere, politics and political action cannot be connected with environmental action for the solution of environmental problems, ignoring that other and different systems and normative and social orders exist outside politics in modern societies that influence and are part of societal transformation. In difference to such "politicist agendas" of global environmental governance, more critical views of environmental action and governance could be described as "trans-political" where other spheres of social action are included in governance, and further scientific knowledge, especially from interdisciplinary research on nature and society, political economy, or human, social, and political ecology.

This critical thinking differs from arguments that environmental governance and action should be only or mainly build on natural scientific knowledge, looking for knowledge as well from natural as form social scientific research that is necessary for identifying possible transformation paths towards sustainability. Young (2011) has argued for more natural scientific knowledge, however, this tends to become an "either-or" question that does not argue sufficiently to answer questions as: which knowledge is required for which problem, in which interdisciplinary perspective? The search of adequate knowledge for environmental governance is often simplified with the reasoning that problems in nature need to be analysed by natural scientists, also policies need to be based on the natural-scientific knowledge. This reasoning is sometimes reduced further to using only knowledge from empirical policy research, where knowledge input is studied for different phases of policy cycles, formulation, implementation and evaluation (Campbell Keller 2009). In the studies of the use of scientific knowledge in international environmental policy by Harrison and Bryner (2004) and Kanie et al. (2014), problems of knowledge use are reflected further, but not sufficiently addressing questions of social-scientific knowledge and of interdisciplinary knowledge integration, taking up this question only in single case studies. This shows that knowledge requirements badly reflected in policy and environmental research. Also in the attempt by Sprinz (2009) to formulate a research agenda for long-term environmental policy, the problems appear in a narrow perspective of policy analysis that does not open for interdisciplinary knowledge use.

Social–ecological transformation to sustainability differs in complicated ways from earlier confrontations, cleavages and conflict lines such as apologetic defence of the existing social and normative orders and reformist changes or radical, utopian or revolutionary critique and combatting of the existing orders and the capitalist system. Dissolving from the scientifically and politically failing project of social emancipation in communism in the twentieth century does not necessarily imply to give up the ideas of social emancipation and social and environmental justice in favour of accepting the political ideology of "there is no alternative" that came with the neoliberal project of economic reforms, or with its scientific justifications (Fukuyama). Rather the alternative is, to learn anew, through analyses of the present social and environmental problems, existing social orders, and the systemic interaction of society and nature in late modernity, how to solve social and environmental problems, to reformulate social emancipation strategies, and to develop social subjects and agency for transformation processes.

In the complicated processes of transition to sustainability appear many ideas that can be seen as creative contradictions and impossibility theorems. Sustainability and degrowth are examples, for such ideas: appearing as impossible in the capitalist system, but necessary in the transition to another economic system. Their "experimental niches" are often strategies of community-based resource management, local economies, alternative forms of life and consumption, that is: utopian islands in the growth-based economy, but necessary as experiments to find ways towards a future sustainable society.

4.4 Discussion and Conclusions—Limits of Global Agency and Global Governance

Five decades of agenda building and policy reforms have shown a series of difficulties in developing global environmental governance. Global governance, when discussed in general (Whitman 2009), without specifying concrete policy fields like environmental governance, or regimes, does not provide the necessary concrete knowledge to develop global environmental governance. More than reflecting principles, norms and power-relations it is necessary to reflect the practices, the ways of policy learning and knowledge use that are visible in the discourse about global environmental governance. With that broadening of the discourse the focus theme becomes more that of agency—how to create global agency and transformative agency in the processes of global environmental governance, where simultaneously with the implementation of international regimes also new forms of collective action and transformative action need to be created. About these forms of transformative action little is known form research, they develop mainly through experimenting, learning and capacity building in the governance processes.

Building global agency—where to begin and how: Sonnenfeld (2006) discusses knowledge generation and application as processes where the "violence of abstraction" prevails, to show the difficulties in analysing and improving global environmental governance. Difficulties result from the complexity and the dynamics in a governance field, where manifold actors, institutions, movements, subjects of interest, asymmetric power relations, and different forms of knowledge production and use interact and clash. Simplified dualisms of the kind "local and global" thinking and action or "small is beautiful" guide many environmental movements; more sophisticated dualisms are constructed in anthropological research where the social and cultural specificity of social and ethnic groups and their livelihoods are endangered through globalising processes and institutions, economic and political forces. With such views transformation towards sustainability seems badly informed. New political dynamics coming with new nongovernmental actors in policy processes, with state- and nonstate governance, open chances of improving governance through experimenting in the sense of adaptive governance (policies as experiments). Global environmental politics, so Sonnenfeld, require abstraction from the specificity of the local and finding paths for the development of global action capacity or agency. In global governance the specificity of sub-global processes and activities cannot be ignored, but the greatest difficulties are such of integrating knowledge and interests to achieve successful global action—abstraction seems to imply "unavoidable oversimplification", and with such simplification comes selectivity in knowledge practices and with regard to the actors to be involved—who shall participate becomes a critical question.

In the field of international environmental policy analysis, the relations and the interaction between national and international policies (Barry and Frankland 2002) show the complexity of clashing interests and ideologies, of asymmetrical power relations, and of coordinating and integrating policies across local, national and global scales. The problems of building a political world order and global agency are studied in different forms and perspectives in governance research: as fragmentation of the architecture of global governance (Biermann 2009; Zelli and van Asselt 2013); as the possibility of a world environmental organisation (Biermann and Bauer 2005); as contrasting forms of state and nonstate governance; as unequal power relations and different forms of ownership and control of natural resources at global, national and local levels. What has been learned from the past experience?

The global environmental awakening happened about fifty years ago, in the late 1960s, when national and international activities began in Western countries to deal with environmental problems that turned out to be global problems, although they were not always perceived and reflected in global dimensions. "Limits to growth" in 1972 was the first hallmark of global environmental governance, discussing the problems in terms of exponential growth of population, natural resource use and economic growth. Subsequent signposts were the sustainability discourse, the conceptualisation of global change processes in terms of anthropogenic climate change, biodiversity loss and land use change, the beginning of global environmental policies with international conventions and regimes as that of climate policy (Kyoto and Montreal Protocols), sustainable development (Agenda 21), biodiversity (Convention on Biological Diversity), and a series of more specific international programmes for global water management, ocean management, food security, desertification.

A series of global conferences in the past five decades symbolised the political adoption of global problems—the 1972 UN-Conference on the Human Environment (UNCHED; Stockholm), the 1974 World Food Conference, the 1992 UN-Conference on Environment and Development (UNCED, Rio de Janeiro), the 2002 World Summit on Sustainable Development (WSSD, Johannesburg), the 2012

Rio+20-Conference on Sustainable Development where important conferences. The main functions of these mega-conferences can be described as setting global agendas, facilitating "joined-up" thinking, endorsing common principles, providing global leadership, building institutional capacity, legitimising global governance through inclusivity (Seyfang and Jordan 2002). The conferences signalise awareness of the problems and efforts to solutions, but contribute little to the building of global environmental governance which has to deal with the complexity and interconnectedness of problems, so that the limits of global governance (Whitman 2009) became a dominant theme. In the practice of global governance the attempts of problem solving dissolve in complicated multi-scale processes of networking and coordination, in which conflicts between actors, institutions and countries are unavoidable. The environmental problems are on the global agenda together with other urgent social, political and economic issues, and the international institutions or regimes that deal with the environment are not always constructed and targeted for that purpose or to prioritise these problems. In the global arena, clash different international regimes, especially that of the economic world order and the climate and sustainability regimes.

Most of the global policy documents and action programmes resulting from the international conferences and governmental meetings are simple declarations of the necessity and the intentions to act, indicating symbolic consensus, without thorough analyses of the problems and potential solutions. Inadequate and insufficient forms of learning from the past policies can be seen in exemplary form in two recent documents—the green economy document of the Rio+20 conference, and the UN/UNESCO-document for the post-2015 UN-development agenda (see Chapter 9). Thirty years after the start of sustainable development initiatives, and in spite of an institutional integration of sustainability in national and international policies, there is not much progress and success in global environmental governance, and the policy documents become more and more wishing lists and exercises in the political rhetoric of appealing to "our common future". The exception of climate policy with its relative success in building an international climate regime cannot justify the deficits of the sustainability strategy. Alternatives to global governance, attempts to bypass the difficulties with the policy and governance processes are not promising: global engineering or geoengineering and genetic engineering become risky ideas and strategies. They are more selective and risky than the policy processes and international regimes, less controlled and more doubtful with regard to solving complex environmental problems such as climate change.

References

Acheson, J. M. (2006). Institutional Failures in Resource Management. *Annual Review of Anthropology, 35*(1), 117–134.

Agrawal, A. (2003). Sustainable Governance of Common Pool Resources. *Annual Review of Anthropology, 32,* 243–262.

Barry, J., & Frankland, E. G. (Eds.). (2002). *International Encyclopedia of Environmental Politics.* London: Routledge.

Bauer, S., & Biermann, F. (2005). The Debate on a World Environment Organization: An Introduction. In F. Biermann & S. Bauer (Eds.), *A World Environment Organization: Solution or Threat for Effective International Environmental Governance?* (pp. 1–23). Aldershot, UK: Ashgate.

Becker, D. C., & Ostrom, E. (1995). Human Ecology and Resource Sustainability: The Importance of Institutional Diversity. *Annual Review of Ecology and Systematics, 26,* 113–133.

Benington, J., & Moore, M. H. (Eds.). (2011). *Public Value: Theory and Practice.* New York: Palgrave Macmillan.

Biermann, F. (2007). 'Earth System Governance' as a Crosscutting Theme of Global Change Research. *Global Environmental Change, 17*(3–4), 326–337.

Biermann, F. (2014). *Earth System Governance: World Politics in the Anthropocene.* Cambridge, MA: MIT Press.

Biermann, F., & Gupta, A. (2011). Accountability and Legitimacy in Earth System Governance: A Research Framework. *Ecological Economics, 70*(11), 1856–1864.

Biermann, F., & Pattberg, P. (2012). *Global Environmental Governance Reconsidered.* Cambridge: MIT Press.

Biermann, F., et al. (2009). The Fragmentation of Global Governance Architectures. *Global Environmental Politics, 9*(4), 14–40.

Bryant, R. L. (Ed.). (2015). *The International Handbook of Political Ecology.* Cheltenham, UK and Northampton, MA: Edward Elgar.

Campbell Keller, A. (2009). *Science in Environmental Policy: The Policies of Giving Objective Advice*. Cambridge, MA: MIT-Press.

Chase Dunn, C. (2013). Five Linked Crises in the Contemporary World System. *Journal of World Systems Research, 19*(2), 175–180.

Daly, H. (Ed.). (1973). *Toward a Steady-State Economy*. San Francisco: Freeman.

Dauvergne, P. (Ed.). (2012). *Handbook of Global Environmental Politics* (2nd ed.). Cheltenham, UK and Northampton, MA: Edward Elgar.

Delanty, G. (Ed.). (2006). *Handbook of Contemporary European Social Theory*. London and New York: Routledge.

Folke, C. (2006). Resilience: The Emergence of a Perspective for Social–Ecological Systems Analysis. *Global Environmental Change, 16,* 253–267.

Forsyth, T. (2003). *Critical Political Ecology: The Politics of Environmental Science*. London: Routledge.

Funtowicz, S. O., & Ravetz, J. E. (1993). Science for the Post-normal Age. *Futures, 25*(7), 739–755.

Gross, M., & Heinrichs, H. (Eds.). (2010). *Environmental Sociology: European Perspectives and Interdisciplinary Challenges*. Dordrecht et al.: Springer.

Gulbenkian Commission on the Restructuring of the Social Sciences. (1996). *Opening the Social Sciences*. Stanford, CA: Stanford University Press.

Hannigan, J. (2006). *Environmental Sociology* (2nd ed.). London and New York: Routledge.

Harrison, N. E., & Bryner, G. C. (Eds.). (2004). *Science and Politics in the International Environment*. Lanham, MD: Rowman and Littlefield.

Jakobsen, S. (1999). International Relations and Global Environmental Change. *Cooperation and Conflict, 34,* 205–236.

Jasanoff, S., & Martello, M. L. (Eds.). (2004). *Earthly Politics: Local and Global in Environmental Governance*. Cambridge, MA and London, UK: The MIT Press.

Jennex, M. E., red. (2014). *Knowledge Discovery, Transfer, and Management in the Information Age*. Hershey, PA: IGI Global.

Kanie, N., Andresen, S., & Haas, P. M. (Eds.). (2014). *Improving Global Environmental Governance: Best Practices for Architecture and Agency*. London and New York: Routledge.

Kanie, N., & Biermann, F. (Eds.). (2017). *Governing Through Goals: Sustainable Development Goals as Governance Innovation*. Cambridge, MA: MIT Press.

Lee, R. E. (2000). The Structures of Knowledge and the Future of the Social Sciences: Two Postulates, Two Propositions and a Closing Remark. *Journal of World-System Research, VI*(3), 786–796.

Lemos, M. C., & Agrawal, A. (2006). Environmental Governance. *Annual Review of Environment and Resources, 31,* 297–325.

Marshall, B. K., & Goldstein, W. S. (2006). Managing the Environmental Legitimation Crisis. *Organization & Environment, 19*(2), 214–232.

Meyer, J. (2001). *Political Nature: Environmentalism and the Interpretation of Western Thought.* Cambridge: MIT Press.

Moore, J. W. (2011). Ecology, Capital, and the Nature of Our Times: Accumulation & Crisis in the Capitalist World-Ecology. *Journal of World-Systems Research, 17*(1), 107–146.

Morin, E., & Kern, A. B. (1999). *Homeland Earth: A Manifesto for the New Millennium.* Creskill, NJ: Hampton Press.

Najam, A., Papa, M., & Taiyab, N. (2006). *Global Environmental Governance: A Reform Agenda.* Winnipeg: International Institute for Sustainable Development.

Nowotny, H., Scott, P., & Gibbons, M. (2001). *Re-thinking Science, Knowledge and the Public in an Age of Uncertainty.* Cambridge: Polity Press.

Ophuls, W. (1977). *Ecology and the Politics of Scarcity.* San Francisco: Freeman.

Pahl-Wostl, C., Giupponi, C., Richards, K., Binder, C., de Sherbinin, A., Sprinz, D., et al. (2013). Transition Towards a New Global Change Science: Requirements for Methodologies, Methods, Data and Knowledge. *Environmental Science & Policy, 28,* 36–47.

Pattberg, P., & Zelli, F. (Eds.). (2015). *Encyclopedia of Global Environmental Governance and Politics.* Cheltenham, UK and Northampton, MA: Edward Elgar.

Pelayo, G. (2008). Environmental Governance and Managing the Earth. Documentary Base. Paris. FNWG. Retrieved from http://www.world-governance.org/spip?article 380. Accessed 11 January 2018.

Peters, R. H. (1991). *A Critique of Ecology.* Cambridge: Cambridge University Press.

Piattoni, S. (2009). Multi-level Governance: A Historical and Conceptual Analysis. *European Integration, 31*(2), 163–180.

Sagoff, M. (2008). *The Economy of the Earth: Philosophy, Law, and the Environment* (2nd ed.). Cambridge: Cambridge University Press.

Seyfang, G., & Jordan, A. (2002). The Johannesburg Summit and Sustainable Development: How Effective Are Environmental Conferences? In O. S. Stokke & Ø. B. Thomessen (Eds.), *Yearbook of International Co-operation on Environment and Development 2002/2003* (pp. 19–39). London: Earthscan.

Sonnenfeld, D. A. (2006). The Violence of Abstraction: Globalization and the Politics of Place. *Global Environmental Politics, 6*(2), 112–117.

Sprinz, D. (2009). Long-term Environmental Policy: Definition, Knowledge, Future Research. *Global Environmental Politics, 9*(3), 1–8.

Strang, V. (2009). Integrating the Social and Natural Sciences in Environmental Research: A Discussion Paper. *Environment, Development, Sustainability, 11,* 1–18.

Swedborg, R. (2014). *The Art of Social Theory*. Princeton: Princeton University Press.

Walker, P. A. (2005). Political Ecology: Where Is the Ecology? *Progress in Human Geography, 29*(1), 73–82.

Wallerstein, I. (2000). Globalization or the Age of Transition? A Long-Term View of the Trajectory of the World System. *International Sociology, 15*(2), 249–265.

Whitman, J. (2009). *The Fundamentals of Global Governance*. Basingstoke: Palgrave Macmillan.

Young, O. R. (2011). Effectiveness of International Environmental Regimes: Existing Knowledge, Cutting-Edge Themes, and Research Strategies. *PNAS, 108*(50), 19853–19860.

Zelli, F., & van Asselt, H. (2013). The Institutional Fragmentation of Global Environmental Governance: Causes, Consequences and Management. *Global Environmental Politics, 13*(3), 1–13.

Part II

Knowledge for Renewing Environmental Governance

5

Environmental Research and Governance: Institutional Problems of Bridging Knowledge Divides and Communicating Science

The discussion of bridging or linking science and practice, science and the public, science and policy, or science and society for solving environmental problems in times of crises, shrinking economies, degradation of ecosystems and reduction of their functions and services (Braat et al. 2012: 13) revealed as first necessity of improving environmental governance to address the weak forms of knowledge integration within and between the practices of production, dissemination and application of scientific knowledge that develops through environmental research. The practices of knowledge integration discussed in this and the following chapter refer to two forms of knowledge creation beyond the dominant form of empirical research:

- *bridging of knowledge divides* between science and practice of environmental governance; this is a part of the broader process of transdisciplinary knowledge production and utilisation, dealing with inter-institutional problems of knowledge integration and communication (this chapter);
- *integrating scientific knowledge from different disciplines and specialised fields of research through interdisciplinary synthesis*; this is an intra-scientific knowledge process dealing with the epistemological and

© The Author(s) 2019
K. Bruckmeier, *Global Environmental Governance*,
https://doi.org/10.1007/978-3-319-98110-9_5

methodological difficulties of integrating knowledge from different specialised disciplines, empirical and theoretical knowledge (Chapter 6).

Knowledge integration happens in multi-scale processes: global environmental governance and transformation to sustainability are themselves paradigmatic examples of multi-scale processes, not only international communication and political action. For all forms of knowledge bridging across the science–practice divide, the common denominator is the assumption that knowledge transfer, communication, exchange, sharing and application are social processes in which different expectations, interests and knowledge forms meet and need to be balanced. Social epistemology as an umbrella term for such and similar forms of collective knowledge practices (Goldman and Blanchard 2018) is not used here for the discussion of knowledge bridging; it is, with all heterogeneity of its approaches, not consequently oriented to transdisciplinary practices of working with knowledge and knowledge integration practically in governance contexts practically. Social epistemology is part of a philosophical discourse of broadening the analyses of knowledge creation and use from individualistic to social forms, as that of Goldman and Blanchard, or referring to practices in scientific knowledge production and use. For the following discussion, it is sufficient to work with the terms of trans- and interdisciplinary knowledge practices; with these terms, the knowledge problems have been discussed for a new global change science (Pahl-Wostl et al. 2013) and in the social–ecological transformation discourse (Hummel et al. 2017).

With the increase of inter- and trans-disciplinary knowledge production and application develop new methods for bridging different social and institutional knowledge practices and integration of scientific knowledge. As the discussion of Pahl-Wostl et al. (2013) and similar ones as that of the International Social Science Council (2010) on transformation processes and transformative action groups show, the methods are pragmatic, not sophisticated methods of research, but methods of collaboration and joint discussion of knowledge requirements: different forms of interviews, group discussions and workshops, participatory methods, and for transition management especially web-based methods, satellite-based data programmes and geographical information systems, scenario-based methods, case studies and group-based modelling. According to these examples show much of the methods used develop through adaptation and modification of already existing scientific methods, using these for specific purposes, although the necessity to develop new and innovative tools and methods is a constant expectation and aspiration.

Knowledge bridging and communication need to deal with the conditions of knowledge transfer and application: the institutional and social conditions of knowledge bridging and integration, the relations between environmental research and environmental action in spheres of social action that are organised differently—science, policy and governance, practices of resource use and management, economic activities, collective action of social movements, lifestyles and resource consumption of different social groups. In bridging the social spheres and knowledge forms two difficulties appear: the specialisation of scientists that limits the problem-solving capacity of their knowledge (Miller and Morris 1999; Kinzig 2001; the broader discussion in human ecology), and the different knowledge practices of decision-makers, resource managers and other actors, not always clear in the demands of knowledge from science (Roux et al. 2006).

Much of the knowledge bridging in environmental research, environmental governance and natural resource management happens still in conventional forms of knowledge transfer and science communication, by use of mass-media, popularising and translating scientific research for non-scientists in society, decision-makers, resource managers, political activists, or the public at large, in attempts to make scientific research applicable or to improve the receptivity of society for science. This limited perspective unidirectional knowledge flow from science as a socially privileged form of knowledge production to vaguely described "practices" of knowledge use and user communities needs, in the further development of environmental governance, be reflected critically; the building of transformative agency requires more knowledge sharing and reflexive processes of knowledge use.

Knowledge bridging and integration require clear distinctions between knowledge processes or practices and other forms of social processes and action; knowledge is part of all forms of social action, but not all of them are reflexive or deliberative knowledge processes. Bridging knowledge forms and knowledge practices—between disciplines, between scientific and other knowledge forms—implies specific forms of communication, mutual understanding and collective learning. Social and ecological systems, mainly coupled through forms of natural resource use in material and energetic processes and resource flows, are not directly affected through the knowledge bridging. The critique of dualistic thinking, the use of hybrid concepts and the levelling of the conceptual differences between nature and society through symbolic communication does not yet lift the barriers between social and ecosystems. Knowledge bridging and integration are only part of reconnecting society and nature, but changes of the coupled systems happen in longer and more complex transformation processes. With all couplings between social and ecological systems, their functions and systemic processes differ: functions of social systems and their services cannot replace functions of ecosystems and their services, and the other way round. The ecosystem services of purification of contaminated air, water and soils cannot be replaced through technologies of purifying and cleaning water, air and soils as part of ecosystem restauration or special economic services. The environmental damages through the large quantities of household and industrial waste cannot be completely solved through technologies and human activities, require the "work of nature" in form of ecosystem functions and services.

5.1   Cognitive Processes and Problems of Bridging Knowledge

The knowledge links between different social spheres of knowledge practices, for example, between science and policy or natural resource management, are usually discussed in terms of application and transfer of knowledge in other, non-scientific forms and spheres of social action.

However, the greater difficulty to organise such forms of knowledge transfer is the selectivity of knowledge use. There is no consensus among scientists about the knowledge to be transferred and used in governance practices; scientific controversies expand into political controversies, as was often observed in environmental research and policy (Sarewitz 2004). Different and competing theories, approaches methods to generate incompatible and contradicting forms of scientific knowledge; knowledge bridging and integration include manifold choices between different and competing forms of scientific knowledge and explanation. Scientific knowledge is communicated and transferred in a "pluralistic knowledge society" (Heinrichs 2005). The selectivity of knowledge communication and transfer pervades all processes of knowledge bridging and integration, evoking questions of the kind: what are the criteria of selection of scientific knowledge that is applied in the political and practice forms of knowledge use? What makes one scientific community more accepted in policy advice than another?

Beyond the selectivity regarding the choice of knowledge and information, the communication processes, the terminology, and the media of communication need to be analysed critically. This is mainly discussed with regard to finding a common language or meta-language to transfer and communicate scientific knowledge, a general problem in interdisciplinary research, where a common language beyond the differing disciplinary terminologies is required to create mutual understanding. The process of science communication is, however, a mediated process with mediating persons (for example, journalists), specific media (mass media), printed and oral or audio–visual communication. One important question in that communication is that of the norms of reporting about science in mass media that follows more journalistic than scientific norms (Dunwoody in Bucchi and Trench 2008: 19ff.).

In global environmental governance, paradigmatically in climate governance, the bridging, linking and communication of science to policy is always controversial and complicated because of such selectivity. The example of climate governance shows that the IPCC, an important institution of knowledge bridging, acts on contradicting premises; it claims to communicate only knowledge about which exists consensus in the community of climate researchers, knowledge which is not selective, and free from political influence or pressures (a conventional view of "objective", value-neutral scientific knowledge). But the members of the panel are nominated by governments. The governmental choices are not only based on the criteria of nominating policy-independent experts (with criteria as credibility); in selecting the experts different and valuing criteria, such as legitimacy or trust and expectations in the experts can be used; the process is transparent in all countries participating in global climate policy. The nomination of experts is part of climate diplomacy and lobbying. In the science–policy communication different and competing communities and networks are active—epistemic communities, policy communities, advocacy groups, coalitions of different interest groups that differ in their goals and interests.

Evident for the knowledge communication in the IPCC-reports is, that the knowledge of a dominant community of natural-scientific climate researchers is communicated to the public and to decision-makers as knowledge about anthropogenic climate change that is created in the conventional view where scientific knowledge is seen as universally valid knowledge; this universal knowledge should also be applied in a global policy process that does not require to take into account different climate narratives or scientific constructions of climate change that are found in the scientific discourse. The pluralism of scientific climate narratives (influenced through the cultural relativising and fragmenting of knowledge processes in such approaches as postmodernism, post-structuralism, social constructivism, political ecology) shrinks to a scientific consensus in terms of probabilities and hard facts Even when these facts are accepted as a basis for communicating climate science, there remain the significant problems how to deal with the different scientific climate change explanations and narratives.

The problems with science communication became evident with the postmodernist discourse:

1. Latour (1993) diagnosed the present situation in science as knowledge fragmented through disciplinary specialisation and conceptual distinctions that can only be found in modernity seen less as historical, more as an ideological concept. Reconnecting knowledge from

disciplinary fragmentation at ideological, ontological, epistemological and methodological levels happens in different forms with different consequences. Latour has only the suggestion, to bridge the disciplinary divide at another level, through a paradigm change in scientific thinking that questions the discourse of modernity, renouncing to distinctions as that between nature and society or culture that have already been blurred through hybrids where politics, science, technology and nature are mixed. In this diagnosis of the knowledge problem, the differences between knowledge production and use and other forms of social relations and action are levelled. Formerly meaningful distinctions are levelled down and the conceptual constructions are no longer clarified but simplified. The hybrids he describes originate from qualitatively different social and ecological systems, from different knowledge forms as the social and the natural sciences. An epistemological bridging process through a new paradigm seems to annihilate more than to integrate the differences, although Latour tries to avoid the pitfalls of postmodernist relativism and constructivism and differentiates knowledge bridging operations into ideological, ontological, epistemological and methodological ones.

Regarding the relationships between nature and society, natural-scientific and social-scientific knowledge, more proposals than that by Latour are required to connect the concepts and bridge the disciplinary knowledge divide. The construction of hybrid concepts such as "socionatures" creates similar difficulties and unclear consequences for knowledge practices as other attempts of constructing unified knowledge practices to integrate science and policy or practice. Whether this merging of concepts is already creating a common language and common ground for the integration of knowledge from different disciplines and knowledge practices can be doubted.

2. The approaches of knowledge bridging can be ranged in a spectrum of variations between the two extremes of conceptual integration at the one end and integration of empirical data at the other.

   For the first form, Wilson (1998) gives an example with the attempt to create unity of knowledge by giving up the distinctions between nature and humans, nature and society as, in his view, dangerous ideological concepts from modernity. But with this attempt not only a conceptual distinction is given up; the bigger problem is that of analysing environmental destruction through material practices of natural resource use. Wilson simplifies the causes and consequences of environmental disruption, reduced them to an ideology of modernity that is replaced through another ideology of unification of knowledge, called consilience in which humans and nature are unified. What happens in fact with the idea of consilience is: in a masked form biological knowledge is used as the paramount knowledge of science and for a unification of science, constructing an ontological unity which ignores the differences between scientific disciplines in producing knowledge relevant for environmental action and governance; as a consequence also knowledge syntheses and the methodological difficulties of these are ignored.

The minimal requirement of using the contested term "ideology" for knowledge production and knowledge use is: not to distinguish between ideological and scientific knowledge in terms of value or interest-based ideologies and value-free or objective scientific knowledge, but to see and analyse the forms of knowledge production and use in terms of power relations and vested interest of social actors and groups. In the interest of whom, for the benefit of whom is scientific knowledge produced and environmental policy done? What are the—sometimes hidden, sometimes openly discussed—interests and values that guide knowledge production and use? Wilson gives another normative view of modernity, with similar consequences as that of Latour: to criticise the distinction between nature and society, social and ecosystems with the suspicion of being ideological. In both cases the concept of ideology is not reflected further, used in a simplified form—and replaced by another, scientifically based, ideology. In Latour's thinking and in actor–network theory, this ideology unfolds in new metaphors (as "the parliament of things") and new hybrid concepts (as "actants"). In the case of Wilson, the ideology is a normative, ethical view of science which unfolds in the concept of "consilience" that has been used in the ecological discourse several times.

The second form of integration mentioned above, empirical knowledge integration, works less with unifying concepts; different examples can be found, where integration of knowledge is an inductive process of puzzling out how to connect empirical knowledge from different disciplines. This happens often in methodologically unclear operations: in forms of moderate, partial and flexible integration without ontological or epistemological constructions of a unity of knowledge. An example for moderate, empirically based knowledge integration gives Nuijten (2011), constructing knowledge integration at the level of interdisciplinary research projects with natural and social scientific components. In this case, it is not assumed that natural and social scientific research are closely connected to specific paradigms, epistemologies, theories or worldviews, but allow for lose integration of knowledge with the help of an integrating framework, connecting to the epistemology of critical realism. This approach seems to deal better with the differences between theoretical, empirical and normative knowledge that cannot be integrated in one continuum of scientific knowledge, but require specific forms of connection that are hardly reflected in epistemology or methodology. Although the integrative framework of Nuijten is only constructed for the use of social and natural scientific research in agricultural science, it can be used for other themes too; the framework works with a differentiated set of classifications of knowledge forms and methodological rules of research that make integration of knowledge a more systematic but cumbersome and complicated process, without ontological and ideological constructions of a unit science; metaphorically it can be described as a "patchwork" of knowledge.

3. Between the two poles of strong and weak integration, dissolution of conceptual distinctions and "patchwork synthesis", exist various variants of knowledge bridging and integration that try to bridge disciplinary and other knowledge divides through epistemological and methodological reflection. An example of conceptual bridging of social and natural scientific knowledge is that of coevolution and co-evolutionary development (Gual and Norgaard 2008). In this case, knowledge integration happens without assuming a similarity between social and ecological systems, or assuming their naturally given integration. Co-evolution of different system types is a framing concept for analysing interactions between the systems. The conditions of development for the co-evolving social and ecological systems need to be specified with regard to different contexts, functions,

structures, processes, temporal and spatial dimensions of development, and differences between sociocultural evolution and biological evolution. Applying the term of coevolution in environmental governance, to analyse the forms and degrees of coupling between social and ecological systems and formulating conditions for their closer coupling does not require dissolving the distinction between social and ecological systems and processes. Identifying and analysing the qualitative differences between social systems and ecological systems is a prerequisite to develop concepts and forms for reconnecting both system types in terms of sustainability or coevolution.

Another, transdisciplinary, variant of "unity with differences" of scientific and practical knowledge is the knowledge bridging through knowledge sharing, where the exchange and flow of knowledge between different social knowledge practices and groups of knowledge bearers, researchers, policymakers, and resource managers, need to be analysed. Such knowledge bridging happens less through mediating concepts, but in discursive and pragmatic forms of discussing and evaluating differences between research and practical knowledge to understand their differences and find out, whether and how they complement each other or can be used in parallel. The knowledge integration can be improved through transdisciplinary research, co-production of knowledge and collaborative learning (Roux et al. 2006).

4. With regard to bridging different knowledge forms and spheres, bridging processes can be classified into three kinds:

(a) *Bridging scientific divides and communicating science to the public* (classified in Table 5.1) implies various forms of dealing with the fragmentation of knowledge production in disciplines, epistemologies and ontologies. Here it is necessary to differentiate between heterogeneous components of scientific knowledge that require bridging—theoretical, empirical and normative knowledge. The forms of communicating science to the public are done under the premise, that it is scientific knowledge that is valid and needs to be understood by non-scientific knowledge users. The process of knowledge bridging appears similar as in education processes; it is a form of training and educating non-scientific knowledge users to make adequate use of scientific knowledge, for example in climate policy, where climate research has become an important theme in science communication for purposes of global environmental governance.

(b) *Bridging scientific and non-scientific knowledge* comes was in the past decades intensively discussed up with the ideas of transdisciplinarity and "mode two" knowledge production, attempting to reconnect local and practical knowledge with scientific knowledge and including them in scientific, for example in participative research. The great divide between scientific and practical knowledge (and similar forms: local, experience-based, tacit knowledge) as a consequence of the differentiation of socially relevant knowledge spheres in the long process of societal development culminated in modern science and its quasi-monopoly of producing true knowledge. The marginalisation and devaluation of other knowledge forms that were long-time important in natural resource use-processes, for example, the knowledge of producers in economic primary production (fishery, aquaculture, agriculture, animal husbandry, forestry) is now converted. In the crisis of scientific knowledge production (discussed in Chapter 4) come rediscovery and recognition of non-scientific knowledge and require the assessment of different knowledge forms for purposes of natural resource management or environmental governance.

(c) *Bridging power-based knowledge divides and inequalities* is not only relevant in transdisciplinary knowledge production; it pervades all forms of knowledge production and use. The building of modern society and the development of modern sciences as part of show that science became an authoritative, power-based form of knowledge production, where not only the intellectual power of scientific knowledge producers is important, more the use of science for strengthening the political and economic power of countries, states, national economies in the race for appropriating natural resources. Five hundred years of a global north–south divide through the building of the modern capitalist world system with core and periphery countries, metropoles and colonies, industrialised and developing countries, with asymmetrical economic, political and military power relations, left deep and lasting influences in the institutions of science and research. The bridging of asymmetrical "knowledge power" becomes practically relevant with international policies and global environmental governance, where questions come up as in the case of climate policy: which scientists and which scientific knowledge about climate change are acknowledged in the building of the global climate regime? Different theories and approaches existing within disciplines are ignored in the processes of knowledge transfer and application where political power becomes a decisive factor. The broader bridging happens as well in scientific as in political debates, in controversies about the democratisation and the greening of science (Jamison 2001; Mathisen 2006).

Bridging knowledge processes and practices happen in more diffuse forms of knowledge integration than the scientifically, epistemologically and methodologically reflected forms of knowledge synthesis. Within the general distinction between two types of knowledge integration, various knowledge practices can be identified, summarised in Table 5.1.

Table 5.1 Forms of knowledge integration: knowledge bridging and knowledge synthesis

---

Knowledge bridging as linking different knowledge forms and knowledge use practices—institutional processes:

1. *Knowledge transfer*—the dominant forms of communication, modification and application of scientific knowledge: transforming knowledge into technologies applied in other, non-scientific spheres of social action (policy, economy, culture, the lifeworld)
2. *Science communication*—communicating the results of scientific, for example, environmental research to the public, to decision-makers and non-scientists (Bucchi and Trench 2008; Jamieson et al. 2017), science journalism, public understanding of science, documentation of scientific knowledge
3. *Knowledge interfacing and knowledge sharing*—policy learning (Kowarsch et al. 2016; Garard and Kowarsch 2017); sharing knowledge between different groups of knowledge bearers and users; bridging scientific and local ecological knowledge in natural resource use and policy or governance processes; transdisciplinary and participatory research; "citizen science"; environmental and social impact assessment
4. *Collective and social learning*—public participation and dialogue (Einsiedel in Bucchi and Trench 2008: 173ff.); participatory research and transdisciplinary knowledge practices; formalised and informal education and training processes; development of "transformative literacy"

(continued)

Table 5.1 (continued)

5. *Media of communicating scientific knowledge*—print media (dominant forms), audio-visual media, internet

6. *Practices of communicators of scientific knowledge*—rules, norms and practices of scientists, experts, journalists; political organisations, environmental movements, NGOs (Yearley in Buchchi and Trench 2008: 159ff.)

7. *Ethics of communicating and sharing scientific knowledge*—discourse ethics (Habermas), ethic committees; dealing with power relations in knowledge transfer and science communication

Knowledge synthesis as intra-scientific knowledge integration—epistemic processes:

1. *Data integration/empirical knowledge syntheses*—meta-analysis, integration through conceptual framing

2. *Theoretical and conceptual knowledge syntheses*—theory formulation; generalisation, induction, deduction; transfer and combination of knowledge, concepts and theories between different disciplines

3. *Integration of theoretical, empirical and normative knowledge*—through conceptual frameworks, double codification, concept transfer and creation of new scientific concepts

4. *Reviewing, monitoring, evaluating research and scientific knowledge*—peer review and other forms of knowledge assessment (assessment reports, consensus workshops, scenario analyses)

5. *Global Environmental Assessments* (Garard and Kowarsch 2017) as large-scale syntheses of scientific knowledge for purposes of use in policy and governance processes are a specific case, using and combining different procedures of knowledge integration, as well syntheses of scientific knowledge as knowledge bridging and communication between science and policy. They are relevant sources for global environmental governance (examples: Millennium Ecosystem Assessment; IPCC-reports on climate change; "Global Environment Outlook"-reports edited by the UNEP)

Sources own compilation; sources mentioned in the text

The forms of knowledge bridging and integration differentiated above are confronted with various procedural and methodological difficulties:

(a) All forms of knowledge bridging share a series of critical and problematic practices, procedural problems of selectivity, effectiveness, legitimacy, credibility, validation, trust and funding of bridging and communication processes. These difficulties are not exclusively of epistemological or methodological kind, but include more difficulties of normative knowledge use and the selection, valuation and assessment of knowledge used for specific purposes as global environmental governance.

(b) Intra-scientific forms of knowledge integration and synthesis are epistemologically and methodologically reflected and theoretically codified, but not sufficiently reflected with regard to the social rules and practices of knowledge integration and the normative influences on scientific knowledge practices. Scientific knowledge production is discussed since the seminal publication of Kuhn (1962) in terms of worldviews, paradigms, and the problems of connecting facts and values, but integration and synthesis of knowledge are much less discussed—epistemologically, methodologically and with regard to the problems of connecting facts and values.

For environmental governance the problems of knowledge bridging have been discussed by Roux et al. (Box 5.1) with a new terminology for cooperative knowledge production, based on a critique of scientific specialisation that was discussed since longer time in human ecology and other interdisciplinary subjects, without significant influence on the practices of science communication, knowledge transfer and sharing. With the sustainability discourse develop—slowly—new knowledge cultures and new methodologies and methods (Pahl-Wostl et al. 2013).

Box 5.1 Knowledge bridging across the science–management divide

Discussing the bridging the knowledge divide between science and management for sustainable management of ecosystems Roux et al. (2006) criticise the unidirectional knowledge transfer from experts to users, suggesting co-production of knowledge through collaborative learning as an alternative. The discussion shows in an exemplary way the possibilities and limits of the two approaches of transfer and co-production of knowledge in environmental science and governance.

The problem—"trained incapacity" of specialists and experts (Miller and Morris 1999): "The diffusion of new knowledge would be simple if there were no social and cultural divides between the suppliers and prospective adopters of knowledge; yet there must be some social or cultural differences between them to drive the need for new knowledge ... the more technologically advanced a knowledge supplier and the more technologically deprived a potential adopter, the bigger the scope for introducing new knowledge but the lower the chance that the transfer of knowledge will be successful. Hence, knowledge vendors (e.g. consultants) generally prefer to work with clients that have levels of technological advancement similar to their own, and in so doing tend to work with those clients that least need their help. This phenomenon, where an increase in technological, educational or economic disparity results in a decrease in transfer potential, can occur between individuals, organizations or countries" (Roux et al. 2006: 1).

The reactions: Push- and pull-strategies between scientists or knowledge producers and decision-makers or knowledge users are the dominant practices of bridging the science—management divide. Scientists focus on pushing knowledge across the divide (for example, through involving end-users in knowledge creation and transdisciplinary research, or through improving the credibility of scientists, or through packaging and translating the knowledge for managers). Managers practice pull strategies (for example, through articulating information needs, improving activities of knowledge seeking and filtering). Although these practices seem complementary, they do not always serve well the purposes of governance. Tacit, person-bound knowledge, for example, can be useful, but cannot be easily used in the one-directional transfer processes. The transfer processes are pre-determined through the defined roles and functional differentiation between scientists and managers or practitioners. New practices of knowledge sharing (that need to be developed in policy and governance processes) can help to build new knowledge cultures and practices that create possibilities of collective learning, of developing communities of learning and communities of practice. In this innovation processes of knowledge bridging and integration the forms of inter- and transdisciplinary knowledge production and synthesis become more important as new cultures of knowledge creation beyond research.

Source Miller and Morris (1999), Roux et al. (2006), Millner and Ollivier (2016).

For global environmental governance the relevant forms of knowledge bridging as transdisciplinary practice and knowledge synthesis as interdisciplinary practice include the following ones:

(a) *knowledge bridging processes:* knowledge interfacing and sharing and collective learning, including public dialogues and consensus finding, building of transformative literacy and capacity of different actors, transdisciplinary and participatory research; among the forms of science communication—communication through scientists, experts and environmental activists

(b) *knowledge synthesis processes:* theoretical forms of knowledge synthesis in form of interdisciplinary theories like that of nature–society interaction, and combinations of various theories; integration of empirical, theoretical and normative knowledge for the solution of specific problems like combatting climate change (including participatory research); review and evaluation of research and available scientific knowledge; critical review, interdisciplinary dialogue and consensus building

The components of participation, cooperation, transdisciplinary knowledge integration and consensus building have in the perspectives of global environmental governance and socio-ecological transformation other functions than hitherto discussed in environmental research and politics. So far involvement of stakeholders in decision-making, participation and cooperation were mainly emergency forms, consequence of knowledge deficits and ignorance in science, reacting to perceived limits of knowledge through building of trust and consensus in situations that have been described in paradigmatic form by Funtowicz and Ravetz as "post-normal science": knowledge is always insufficient, decisions are urgently required, and consensus between actors needs to be built. The forms of cooperation and broader participation of actors are necessary for global environmental governance, however, not in that limited perspective of dealing with limits of knowledge: more for experimenting with new forms of knowledge integration and sharing to shift the limits of knowledge that are traditionally seen as limits of knowledge in specialised research that require further research. That new knowledge can be generated in other forms too, in inter- and transdisciplinary knowledge practices, is only gradually recognised: not yet in academic research, mainly in new, interdisciplinary fields of research as in the environmental sciences, where such knowledge integration is practically necessary. In these practices of cooperation changes also the view of participation of non-scientific stakeholders in knowledge production from a solution for dealing with ignorance to a solution for creating new knowledge through knowledge integration.

In this way of knowledge integration through inter- and transdisciplinary knowledge practices and procedures change gradually the views of scientific knowledge, its creation, validation and application—as a consequence of new forms of knowledge production discussed in the past decades. The aim is not creating ideological consensus as, for example, in consilience, or through replacing lack of knowledge with ethics, visions and worldviews, but to develop new knowledge practices in which controversies, conflicts and dealing with competing forms of knowledge becomes a routine. Controversies, conflicts and the necessity of their solutions are not primarily a consequence of lacking knowledge, of complex systems and limits of scientific knowledge, but of the organisation of society and economy, the socio-metabolic regime of industrial society and the capitalist world system with asymmetrical power relations in science, politics and the economy. Contrasting theories and explanations—a usual phenomenon in scientific disciplines where often no coherent knowledge is found and no consensus about different theories and approaches—are not indicators of "bad science" that require more research or can be compensated and mediated through consensus and shared goals. The plurality and the competition of different paradigms and theories are part of the scientific processes of knowledge production and need to be dealt continuously through knowledge bridging and integration for developing environmental agency and transformative capacity. To deal with contested and essentially contested concepts, without always dissolving them or replacing them through new scientific concepts, implies to redefine contested concepts with new knowledge and practical experience, as it happens, for example, with the transformation of the concept and perspective of sustainable development through the interdisciplinary concept of social–ecological transformation.

5.2 Essentially Contested and Contested Concepts

Specific methodological and practical problems appear in the knowledge bridging and integration processes for environmental governance and sustainability transformation through the use of contested ideas and concepts which are always used with different and competing interpretations.

1. *"Essentially contested concepts"—the example of sustainable development.* Essentially contested concepts (Collier 2006), unclear concepts or concepts with multiple and competing meanings, cause special difficulties in bridging and communicating knowledge, as can be illustrated with the notion of sustainable development (Bruckmeier 2016):

- it is denied the status of a scientific concept (although it can be used in science and defined in ways that it can be used in environmental research);
- it has multiple meanings, is interpreted by scientists and political actors in many different ways which can only be understood when the term is operationalised or translated in political strategies;
- the dominant interpretations of the term change in the course of time (after thirty years of public use, the term is now interested anew with the guiding ideas of innovation, transition and transformation of societal systems);
- it is a bridging concept in a specific meaning, providing through its multi-semantic quality the possibility of using it differently in the social practices of environmental communication and political action, but still creating consensus among many persons, social actors, organisations, interest groups that one is acting collectively for a common goal.

There was never scientific or political consensus about the interpretation of sustainable development and the ways to achieve this goal. Nevertheless it became the most successful, until today not replaced idea in environmental governance. Since the global UNCED conference in 1992 it is the overarching concept for all local, national international environmental policies; in the policy processes, it became evident that environmental sustainability needs to be matched with social and economic sustainability to become effective and to achieve compromises between contrasting goals. As a consequence of that the idea of sustainable development has been transformed into a concept that guides environmental policies and is operationalised in a series of indicators without that scientific consensus has been created about the interpretation of the concept in scientific terms. This can also be described differently: the idea of sustainable development bypassed the scientific discussion and clarification, was operationalised through political action programmes. In this form temporary stability in the application of the term was achieved, at a low level of clarification. The clarification and redefinition of the term happen in the scientific discourse, parallel to the policy process. In the scientific sustainability discourse two contrasting trends can be observed since some years:

- to criticise the term as an over-complex ecological one that cannot provide meaningful information for policy processes and replacing it by other, simpler concepts;
- to develop the term and redefine it theoretically with the concepts of innovation, transition and transformation, mainly in social-scientific discussions.

Why the idea is fundamentally criticised again after long successful application, has several reasons. Politically it is often seen as an unclear and ageing or outdating concept that needs to be replaced by new ones because it does no longer work as an effective integrating concept in public policies. The slowdown and limited success of sustainability policies added up to the critique. In the scientific discourse about the environment and environmental governance, the critique is growing that sustainability is a term describing hyper-complex processes and states of interacting social and ecological systems that can never be controlled in resource management and public policies because of that complexity (Benson and Craig 2014). In this situation and in search for concepts to replace it, the term of resilience has gained influence in a paradigm competition.

2. *Paradigm competition—resilience and sustainability in interdisciplinary contexts.* Resilience has clear meanings in the analysis of ecosystems (engineering resilience, ecological resilience, social–ecological resilience: Folke 2006; Folke et al. 2016). With its broader use the term (including processes in interacting social and ecological systems, and the reinterpretation of processes of development and change as adaptation to

unforeseeable disturbances), tends to deviate from the original meaning of sustainability as a long term, intergenerational process. With the shifting of the meaning of sustainability towards resilience, the critical debate about sustainable development is interrupted, ignoring the debate about social–ecological transformation of modern society as a long-term process; sustainability appears now as short-term stability of a specific constellation of coupled social–ecological systems that develop with the limited possibilities of adaptive cycles (constructed from ecosystem research), the development more determined through disturbances and the search for new temporary stability after that. Resilience can, as sustainable development, be criticised for its vagueness, its controversial use for describing ecosystem processes, the orientation to unforeseeable disasters, or the ideological nature of its use in the social sciences, but more important seem critical analyses of the practical consequences of using this concept that can be seen in the policy processes. Resilience becomes a form of thinking sustainability in a more short-term perspective of "getting through", surviving and coping with disturbances that cannot be planned for; although different in its meaning it has functional similarities with incrementalism in policy research. The ideas of incrementalism or gradualism in the political sciences (Lindblom 1959) were used to describe, how decision-making works: through small steps and with the logic of "muddling through". Resilience, an ecological concept originally not meant for political processes, can have, unintended or intended consequences of giving up the long-term perspective of systems transformation that is required for sustainability. With the use of resilience the term of sustainability is cut back to adaptation. The model of the adaptive cycle seems to be locked in an ecological process thinking which does not open for a long-term perspective of development and transformation of societal systems. Instead of initiating transformation processes, sustainability becomes in the sense of the metaphor of "muddling through" a form of change that can be described with another metaphor—"balancing at the edge of the abyss" (Ibargüen 1989).

Resilience as value-loaded construction with the connotations of individualisation, self-responsibility and assumptions of a future that cannot be influenced, planned, controlled, tends to make sustainability or sustainable development meaningless (Meyen et al. 2017:

172). When sustainability is reduced to a norm or political goal it is easier to criticise it, especially when it is connected with an inexact formulation of the temporal perspective. The original intentions of sustainability—intra- and inter-generational solidarity, redistribution of resources, solving the overuse and pollution problems in modern society—are intentionally or non-intentionally given up, or modified in resilience thinking. This situation is given meanwhile, when two interpretations of sustainability compete with each other,

- the original one of a long-term perspective of action to create intra- and inter-generational solidarity in resource use, and
- the modification of sustainability and the concept of resilience, that is sometimes resulting in the conclusion to give up the concept of sustainability for the analysis of natural resource use and replace it by the simpler, less demanding concept of resilience (Benson and Craig 2014).

The reduction of sustainability to resilience results in new controversies; it is challenged through the renewal and reinterpretation of sustainable development as socio-ecological transformation as it unfolds in the discourse of earth system governances and in social ecology. Without advancing the perspective and concept of socio-ecological transformation, it seems difficult to achieve the goals of sustainable development.

3. *New science-related ideologies—consilience, citizen science.* The risks and threats of environmental global environmental change are similar for all humans, whereas the solutions through environmental policy are only rhetorically that of "our common interest" and "our common future". So far no adequate forms of political and collective action have been found to achieve the goals of sustainability. Environmental policies are in the interest of specific social groups and countries, not necessarily for the benefit of all; they indicate the national and international power relations and the limits of interests to change the global economic system. The vested interests of powerful actors and global players are sometimes difficult to identify behind the veil of environmental rhetoric. Global public goods for humankind can be created through the sharing and redistribution of natural resources globally, intra- and inter-generationally, and transforming the industrial society towards an environment-friendly

post-industrial society—all ideas connected with sustainable development in the ecological discourse. With the problems of knowledge bridging, integration and synthesis in environmental research also new scientific ideologies develop in attempts to support sustainability, but some of these are leading astray, trying to solve the problems of knowledge integration not empirically as in interdisciplinary knowledge integration, but normatively or through ontological reasoning and formulating new worldviews.

An example for such ontological redefinitions of scientific knowledge production happens with the idea of "consilience" (Wilson 1998), an idea of the unity of knowledge, including scientific and other forms as cultural knowledge in the arts and in religion. With Wilson's concept of consilience re-emerge attempts to create a unity of science formulated earlier in different forms, in the nineteenth century, for example, in the visionary idea of a "universology", a science of the universe in the meaning of a unity of all human knowledge and activities. formulated by the US-American anarchist and philosopher Stephen Pearl Andrews (1872). Wilson himself refers to Snow (1959) and his discussion of two different cultures, of the sciences and the humanities, which split knowledge production in modern society and prevented the solution of problems. The diagnosis is similar to that giving rise to interdisciplinarity, which, however, is not based on the idea of an original unity of knowledge that is lost and can be recreated with the help of normative, ontological or ideological ideas. It does not seem to make sense to think about a reconnection of all specialised knowledge; the forms of knowledge synthesis in interdisciplinary knowledge generation are limited, complementary to the forms of specialisation, working with that specialised knowledge and its integration for specific themes and purposes as that of solving environmental problems.

Wilson begins with the diagnosis that the rapidly increasing differentiation and specialisation of knowledge in modern sciences dissolved the original unity of knowledge production. He seeks the solution in a belief of unity, revealing as a final sense of the sciences, the humanities and the arts a common goal that shows the final truth of all knowledge. He formulates it as the conviction that the world is well organised, based on a few natural laws that are invisible in the mass of data. For consilience, it is not required to do the scrupulous and difficult work of knowledge syntheses, believing in a transcendental reality taken for granted, that there is a final truth in all knowledge makes it simpler. Before Wilson's re-interpretation of the philosophical idea of consilience there have been other attempts in the history of modern environmental movements to recreate a metaphysical unity of science, for example, Capra's (1975) "Tao of physics" based on the idea that physics and Eastern mystics provide a common truth, or in his later attempt (Capra 2002) to show the hidden connections of knowledge through systems and complexity theory. In difference to knowledge integration and synthesis by way of methodologically guided interpretation, such transcendental construction of unity of knowledge are working with ontological, metaphysical, ideological, religious ideas and beliefs.

Other, less transcendental ideologies of knowledge and science work with normative ideas about the production and application of scientific knowledge, for example, attempts to connect science and practice or science and lifeworld accompanied, articulated in the ecological discourse, by environmental movements and in environmental science in the last half century. Among these are new interpretations of the diffuse idea of a "citizen science" as articulated by Irwin (1995) in the idea of scientific citizenship through the opening the sciences and science politics towards the public, which is especially important with regard to environmental problems and complex processes a sustainable development. Citizen science can be understood as an overarching concept to include different forms of reconnecting science and practice in the interest of those who have to share the burdens and the benefits of environmental policies. Similar ideas of opening the social sciences to more transdisciplinary, participatory, policy- and practice-oriented forms of knowledge production boomed towards the end of the twentieth century, with the discourses of new forms of knowledge production, including transdisciplinarity, "mode two", with the ideas in the report of the Gulbenkian Commission on the Restructuring of the Social Sciences (1996) about the opening of the social sciences, and more recently with the report of the International Social Science Council (2010) about transformative science. These ideas take up ideas form the discourse of critical theory, but in simpler and normative forms that do not always support the ideas of societal, economic, or social–ecological transformation.

4. *Metaphors in science, in the production and communication of knowledge.* In the environmental discourse "wild cards" are an important theme, referring to the significance of unexpected or low-probability events in forms of physical and other forms of disturbance for environmental action, policies or governance. The metaphorical notion of "wild card" continues to be used and communicated as metaphor, also when it has meanwhile adopted different meanings as a technical term, in science, in future studies, and in the environmental discourse. Metaphors have multiple functions in scientific knowledge production, dissemination and application, functions which are so far insufficiently reflected, analysed and systematised (Kuhn 1979; in ecology: Pickett and Cardenosso 2002; in sociology: Lüdemann 2004). The discourse about metaphors in philosophy and science is intensifying in the past decades. Traditionally science and philosophy are seen with their cultivation of precise concepts as the opposite of metaphorical thinking. This view has lost significance since the origin of many scientific concepts from metaphors is discussed, even the idea that all scientific thinking is rooted in metaphors. Metaphors that are transformed into scientific concepts have a paradigmatic role in scientific communication: the creation of new concepts through comparison, similarities or analogies (how metaphors are explained since the classical definitions in ancient poetic and rhetoric by Aristotle).

Metaphors in science are forms of creating new scientific concepts. The newer philosophical metaphorology (Blumenberg 1960; Derrida and Moore 1974) identified metaphors as the beginning of scientific thinking. Scientific concepts began as metaphors, for example as concepts transferred from one discipline to another (as the term metabolism from biology to the social sciences) where they, in the course of time, adopted new meaning as theoretical concepts. The broadening of knowledge-generation processes through transgressing disciplinary boundaries can be described in two important forms and phases of knowledge generation: *in knowledge discovery processes*, how new knowledge is created through the use of new metaphors; and *in knowledge synthesis processes*, where metaphors again come into use to codify the integration of knowledge for which new concepts need to be created. In both processes, metaphors help to generate new concepts and channel the flow of knowledge from one discipline to others. Bradie (1999) described three main cognitive functions of metaphors in science: the *rhetorical function* (traditional role of metaphors in pedagogics, in rhetoric), the *heuristic function* (in the knowledge creation process or discovery), including the building of new disciplines (like sociology in the Chicago School developing from ecological metaphors and concept transfer), and the *cognitive or theoretical function* (justification and validation of theories). Since the theory of cognitive metaphors (Lakoff and Johnson 1980) developed a broader discussion about metaphors in different knowledge spheres, in science, politics, education, everyday communication. Biere and Liebert (1997) give an overview of debates of metaphors in science, in science communication and knowledge bridging in which the cognitive functions of metaphors—as creating understanding and communication between separate knowledge practices—is elaborated. This seems to indicate a breakthrough in the discussion of metaphors that have long time been limited to the aesthetic debates about language use in poetics and rhetoric.

The consequences of dealing with unclear or contested concepts and metaphors, with competing paradigms in the ecological discourses in science and politics are contradicting; the concepts can develop creative and innovative capacities, or they can prevent the clarification of concepts; in any case, they have heuristic functions of discovery and creation of new knowledge. Not all the functions of contested concepts and metaphors in scientific knowledge production and application are already clear. However, contested and unclear concepts, or unfinished concepts like metaphors, are more widespread and more important for scientific knowledge production than hitherto thought or admitted. Vague, unclear terms with multiple meanings and continuous interpretation and controversial discussion have obviously manifold forms and functions as bridging concepts, as the few terms of sustainability, resilience, governance discussed here show: knowledge production and application, in environmental science and politics is driven by these unclear or unfinished concepts.

The conventional method to define scientific concepts is captured in the dilemma of explanation through generalisation; most definitions of scientific concepts are abstract and general, need for their application specification and further interpretation. The few methodological rules for definitions, inherited from philosophy and linguistics, and the creation of concepts through classification, typology, taxonomy, show the dependence of concepts from other concepts that are abstract and unclear. Definitions keep concepts in a state of abstraction that they need to be specified. The operational practice of definitions in the natural sciences, to define concepts through the procedures of measuring them, is not a definition in the strict sense, it dissolves concepts; it is also not applicable for all concepts.

Unclear and contested concepts seem to have further qualities that make them attractive in science as well as in the practices of environmental policy and governance. They help to discover new knowledge, to find new ideas and explanations, to deal with problems for which clear concepts are not yet developed. Concepts need to be created in long times of reflection, discussion, analyses, comparisons—in this situation metaphors seem irreplaceable, and a widespread way of creating metaphors for not yet sufficiently investigated phenomena is concept transfer from one discipline to another. Metaphors have another useful quality in search of knowledge and developing concepts—they illustrate a phenomenon in simple form through comparison, analogies, or similarities; all of these cognitive operations are widely used in science. Finally, unclear concepts and metaphors are useful as bridges of thinking and heuristics in creating knowledge about problems that require further research, for which one cannot interrupt the decision-making process until scientific clarification or consensus has been achieved. The paradigmatic example discussed here is the notion of sustainable development: it is a process that stretches into the distant future which is unknown and cannot be made known through empirical research; it can only be detected successively through combinations of different knowledge practices, including research, knowledge synthesis, scenario analysis. Trivial and ironic explanations of the future, widespread and well known, are useless here: Popper's ironic comment that the future is unknown, that is why it is called future; Keynes sophisticated irony "in the long run we are all dead", which justifies practices in economics and economic thinking and action to deal with the short-term future only. A distant horizon of the future is necessary in environmental research and governance, and if no methods and knowledge procedures are available for that purpose in science, it is necessary to create new knowledge through interdisciplinary knowledge synthesis.

5.3 Conclusions: Practices of Bridging and Applying Knowledge

The manifold ideas and forms of knowledge bridging and linking give rise to heterogeneous practices, many of them improvisations, some also creating confusion. Yet, the discussion of bridging knowledge divides with regard to knowledge use for global environmental governance and socio-ecological transformation brought some clarification about the necessity of new or underused forms of knowledge bridging and integration which can be described as follows:

1. *Heuristics* as methods of problem-solving and discovery or creation of new knowledge are methods for problem-solving and suffice the practical purposes of knowledge use in environmental governance processes. They deserve more attention as knowledge providing tools, showing that relevant knowledge for the practices of governance is not only produced through research.
2. The *development of new (theoretical) concepts* is to a large degree one of transforming metaphors into concepts—through concepts transferred from one discipline to another, where they are not yet concepts, but have to be transformed into new concepts. This process of concept creation was up to now mainly discussed in philosophy and in the general debate about metaphors in science is not yet reflected and systematically used in environmental research and governance.
3. Forms of *participatory knowledge production* like knowledge sharing, interfacing and collective learning or deliberative policy learning develop in the use and application of scientific knowledge for global environmental governance, in processes of knowledge bridging,

integration, transfer and science communication. These creative and knowledge-creating processes are underestimated and underused as ways to develop and improve environmental governance.

4. *Collective knowledge syntheses*, synthesis workshops and future workshops are already established and widely used methods for producing knowledge for environmental governance, especially in global assessment projects. Syntheses can be improved through combination with other relevant methods for environmental governance, especially scenario analyses and participatory scenario methods. In scenarios can further methods be used, heuristics, intuitive methods and methods for idea—and solution-finding. Global environmental governance is, however, not a cognitive and knowledge-generation practice according to textbooks and scientific standards or experimenting with methods. Methods applied and combined have clear functions and purposes in relation to the aims of the governance process and the problems to solve. The scientific knowledge input in policy processes not mainly that of creating toolboxes, rather the toolboxes are created in the governance processes.

5. *Policies as experiments*, the idea from adaptive management and sustainability science, has not yet gained sufficient attention and interest in environmental governance practices as a new instrument for knowledge creation when research is not possible or not effective. It needs to be developed and transformed through the use in governance processes, cannot be developed as an abstract idea.

References

Andrews, S. P. (1872). *The Basic Outline of Universology.* New York: D. Thomas.

Benson, M. H., & Craig, R. K. (2014). The End of Sustainability. *Society and Natural Resources, 27,* 777–782.

Biere, B.-U., & Liebert, W.-A. (1997). *Metaphern, Medien, Wissenschaft.* Opladen: Westdeutscher Verlag.

Blumenberg, H. (1960). Paradigmen zu einer Metaphorology. *Archiv für Begriffsgeschichte, 6,* 7–142.

Braat, L. C., & de Groot, R. (2012). The Ecosystems Services Agenda: Bridging the Worlds of Natural Science and Economics, Conservation and Development, and Public and Private Policy. *Ecosystem Services, 1,* 4–15.

Bradie, M. (1999). Science and Metaphor. *Biology and Philosophy, 14,* 159–166.

Bruckmeier, K. (2016). *Social–Ecological Transformation: Reconnecting Society and Nature.* Houndmills, UK: Palgrave Macmillan.

Bucchi, M., & Trench, B. (Eds.). (2008). *Handbook of Public Communication of Science and Technology.* London and New York: Routledge.

Capra, F. (1975). *The Tao of Physics.* Shambhala Publications.

Capra, F. (2002). *The Hidden Connections.* London: HarperCollins.

Collier, S. J. (2006). Global Assemblages. *Theory, Culture & Society, 23*(2–3), 399–401.

Derrida, J., & Moore, C. T. (1974). White Mythology: Metaphor in the Text of Philosophy. *New Literary History, 6*(1), 5–74.

Folke, C. (2006). Resilience: The Emergence of a Perspective for Social–Ecological Systems Analyses. *Global Environmentla Change, 16*(3), 253–267.

Folke, C., Biggs, R., Norström, A. V., Reyers, B., & Rockström, J. (2016). Social–Ecological Resilience and Biosphere-based Sustainability Science. *Ecology and Society, 21*(3), 41. https://doi.org/10.5751/ES-08748-210341.

Garard, J., & Kowarsch, M. (2017). Objectives for Stakeholder Engagement in Global Environmental Assessment. *Sustainability, 9,* 1571. https://doi.org/10.3390/su9091571.

Goldman, A., & Blanchard, T. (2018). Social Epistemology. In E. N. Zalta (Ed.), *The Stanford Encyclopedia of Philosophy* (Summer Edition). https://plato.stanford.edu/archives/sum2018/entries/epistemology-social/.

Gual, M. A., & Norgaards, R. (2008). Bridging Ecological and Social Systems Coevolution: A Review and Proposal. *Ecological Economics, 69*(4), 707–717.

Gulbenkian Commission on the Restructuring of the Social Sciences. (1996). *Opening the Social Sciences.* Stanford, CA: Stanford University Press.

Heinrichs, H. (2005). Advisory Systems in Pluralistic Knowledge Societies: A Criteria-Based Typology to Assess and Optimize Environmental Policy Advice. In S. Maasen & P. Weingart (Eds.), *Democratization of Expertise?* (pp. 41–62). Dordrecht: Springer.

Hummel, D., Jahn, T., Keil, F., Liehr, S., & Stiess, I. (2017). Social Ecology as Critical, Transdisciplinary Science—Conceptualizing, Analyzing and Shaping Societal Relations to Nature. *Sustainability, 9,* 1050. https://doi.org/10.3390/su9071050.

Ibargüen, R. R. (1989). *Narrative Detours: Henry Miller and the Rise of New Critical Modernism*. Ph.D. thesis, Yale University.

International Social Science Council (ISSC). (2010). *Transformative Cornerstones of Social Science Research for Global Change*. Paris. www.world-socialscience.org.

Irwin, A. (1995). *Citizen Science: A Study of People, Expertise and Sustainable Development*. New York: Routledge.

Jamieson, K. H., Kahan, D. M., & Scheufele, D. A. (Eds.). (2017). *The Oxford Handbook of the Science of Science Communication*. Oxford: Oxford University Press.

Jamison, A. (2001). *The Making of Green Knowledge*. Cambridge: Cambridge University Press.

Kinzig, A. P. (2001). Bridging Disciplinary Divides to Address Environmental and Intellectual Challenges. *Ecosystems, 4*, 709–715.

Kowarsch, M., Garard, J., Riousset, P., Lenzi, D., Dorsch, M. J., Knopf, B., et al. (2016). Scientific Assessments to Facilitate Deliberative Policy Learning. *Palgrave Communications 2*, art, 16092. P. C. ISSN 2055–1045 (online). https://doi.org/10.10057/palcomms.2016.92.

Kuhn, T. S. (1962). *The Structure of Scientific Revolutions*. Chicago: University of Chicago Press.

Kuhn, T. S. (1979). Metaphors in Science. In A. Ortony (Ed.), *Metaphor and Thought* (pp. 409–419). Cambridge: Cambridge University Press.

Lakoff, G., & Johnson, M. L. (1980). *Metaphors We Live By*. Chicago: University of Chicago Press.

Latour, B. (1993). *We Have Never Been Modern*. Cambridge, MA: Harvard University Press.

Lindblohm, C. E. (1959). The Science of 'Muddling Through'. *Public Administration Review, 19*(2), 79–88.

Lüdemann, S. (2004). *Metaphern der Gesellschaft: Studien zum soziologischen und politischen Imaginären*. München: W. Fink.

Mathisen, W. C. (2006). Green Utopianism and the Greening of Science and Higher Education. *Organization & Environment, 19*(1), 110–125.

Meyen, M., Karidi, M., Hartmann, S., Weiss, M., & Högl, M. (2017). Der Resilienzdiskurs: Eine Foucault'sche Diskursanalyse. *GAIA, 26*(S1), 166–173.

Miller, W. L., & Morris, L. (1999). *4th Generation R&D—Managing Knowledge, Technology, and Innovation*. New York: Wiley.

Millner, A., & Ollivier, H. (2016). Beliefs, Politics, and Environmental Policy. *Review of Environmental Economics and Policy, 10*(2), 226–244.

Nuijten, E. (2011). Combining the Research Styles of the Natural and Social Sciences in Agricultural Research. *NIJAS-Wageningen Journal of Life Sciences, 57,* 197–205.

Pahl-Wostl, C., Giupponi, C., Richards, K., Binder, C., de Sherbinin, A., Sprinz, D., et al. (2013). Transition Towards a New Global Change Science: Requirements for Methodologies, Methods, Data and Knowledge. *Environmental Science & Policy, 28,* 36–47.

Pickett, S. T. A., & Cardenasso, M. L. (2002). The Ecosystem as a Multidimensional Concept: Meaning, Model, and Metaphor. *Ecosystems, 5,* 1–10.

Roux, D. J., Rogers, K. H., Biggs, H. C., Ashton, P. J., & Sergeant, A. (2006). Bridging the Science-Management Divide: Moving from Unidirectional Knowledge Transfer to Knowledge Interfacing and Sharing. *Ecology and Society, 11,* 34, art 4. http://www.ecologyandsociety/org/vol11/iss1/art4/.

Sarewitz, D. (2004). How Science Makes Environmental Controversies Worse. *Environmental Science & Policy, 7,* 385–403.

Snow, C. P. (2001 (1959)). *The Two Cultures.* London: Cambridge University Press.

Wilson, E. O. (1998). *Consilience: The Unity of Knowledge.* New York: Alfred A. Knopf.

6

Interdisciplinary Knowledge Integration for Environmental Governance: Epistemological Questions

The development of interdisciplinary research and science is a reaction to the limits of disciplinary specialisation as the main form of development of academic science. Forms of interdisciplinary knowledge integration in science and in knowledge transfer for environmental governance include synthesis of empirical knowledge or data and theoretical knowledge synthesis in different forms (see Chapter 5, Table 5.1). The difficulties of integration of knowledge from natural- and social scientific research, experienced in the environmental sciences, can be described as epistemological and methodological problems. In general knowledge integration, and synthesis are epistemologically and methodologically less developed than knowledge production through research. Epistemological questions to be discussed further include the clarification of knowledge integration and synthesis for environmental governance:

- How to connect or integrate knowledge from ontologically, epistemologically and methodologically different disciplines?
- How to connect different theoretical concepts and theories in environmental research and interdisciplinary social ecology?

© The Author(s) 2019
K. Bruckmeier, *Global Environmental Governance*,
https://doi.org/10.1007/978-3-319-98110-9_6

- How to use normative and value-loaded terms, including metaphors, in environmental science and social ecology?

Transdisciplinary knowledge syntheses between scientific and non-scientific knowledge forms imply further epistemological and methodological difficulties:

- How to connect scientific and practical or local knowledge in environmental research and governance?
- How to connect environmental research and governance with ethical reasoning?
- How to integrate knowledge along the chain of generation, dissemination, transfer and application of scientific knowledge in governance processes?

These questions are not yet systematically discussed and answered in interdisciplinary research. Answers develop gradually, with the spreading and the advances of interdisciplinary research. Also, new scientific ideologies (see Chapter 5) that argue for a unity of human knowledge in all its forms need to be reflected epistemologically and with regard to the use and misuse of the ideology concept that resurfaces in debates as that of consilience.

Epistemology is understood here, for the purpose of clarifying interdisciplinary knowledge integration for environmental governance, in a broad sense as a theoretical reflection of forms of generation, synthesis, dissemination and application of scientific knowledge. The epistemological discourse has expanded in the twentieth century into many and competing approaches and theories of science and scientific knowledge production; this is not discussed further here, only with regard to its components of knowledge integration and synthesis. The discourse is not specialised along disciplinary lines, is more general and abstract, although it is difficult to find general rules for scientific knowledge production. The differentiation of the sciences that prevents general rules is that of the divide in humanities, social sciences and natural sciences. The differentiation of the generation of knowledge in the humanities, the social and the natural sciences creates some common rules (as that of verification of all scientific knowledge) and many different ones. The epistemological approaches that work with epistemic models for the natural sciences, positivism and neopositivism, are limiting the epistemological reflection to empirical research as the main way of scientific knowledge production, paying less attention to knowledge synthesis and to theoretical knowledge production that is more important in the social sciences.

Knowledge syntheses across the social and natural sciences that come with environmental research work to a large degree with preliminary methods, showing the epistemic differences between the social and natural sciences and the difficulties to deal with these. Before the expansion of environmental research in the last decades, no significant knowledge exchange, transfer and integration of knowledge happened between the two groups of sciences, except in some synthetic disciplines as anthropology or geography that include natural- and social-scientific knowledge. Together with some other interdisciplinary subjects like human, social and political ecology, these deliver the models for knowledge integration in environmental research that need to be developed further.

The epistemic problems of knowledge synthesis for the environmental sciences include questions of integrating different knowledge forms to deal with in syntheses—scientific, including theoretical and empirical knowledge, normative knowledge, experiential knowledge, local and practical knowledge. Furthermore, questions of limits of knowledge come up in environmental research and research about nature–society interaction (Salmon et al. 1999: 112ff.), regarding general limits and limits at disciplinary levels. With interdisciplinary knowledge integration and synthesis new forms of knowledge creation develop without further empirical research, based on more systematic use of available knowledge that is dispersed in different disciplines and fields of research. Knowledge can be enlarged through synthesis of available knowledge and through further theoretical codification and generalisation in the process of synthesis. Knowledge gaps can be reduced, through transferring concepts and combining empirical and theoretical knowledge, which is usually only done within a discipline. The potential of interdisciplinary research and knowledge integration that has developed rapidly with the increasing environmental research in the past decades is not yet fully recognised in academic research. Interdisciplinary research and knowledge integration are still seen as exceptional forms that apply only for specific problems; with the progressing subdisciplinary specialisation of research knowledge synthesis becomes more necessary and simultaneously more difficult.

Differentiation of disciplines or specialisation of knowledge production is not a well organised process of division of knowledge production; parallel and double research is often found, forms of overlapping research and similar knowledge production in different disciplines. Also, interdisciplinary knowledge production and synthesis is not a coherent forms of knowledge production; it advances in different and competing forms. Interdisciplinary knowledge practices are not yet generally accepted as a way to deal with or shift the limits of specialised knowledge production. In the practices of research knowledge exchange between disciplines is not always reflected in terms of interdisciplinarity.

6.1   The Development of Interdisciplinarity

Academic science and research, characterised through specialisation, reached its modern forms with the development and progressing differentiation of the natural sciences since the seventeenth century, and the development of the social sciences since the nineteenth century. Interdisciplinary research and knowledge synthesis are attempts to solve the problems that come with specialisation in disciplines and subdisciplines, the fragmentation of scientific knowledge and the limits of specialised research in dealing with complex problems. Creating a new specialisation of complexity science, as done with the analysis of complex adaptive systems, is not the only solution. The environmental problems discussed here require as well integration of knowledge from several disciplines as new forms of interdisciplinary research and broader syntheses.

Interdisciplinary research and knowledge syntheses developed rapidly in the second part of the twentieth century, when also environmental movements and environmental research which made use of interdisciplinary knowledge production developed. During the first international conference on interdisciplinary research in 1970 (OECD 1972) forms of borrowing and knowledge exchange were discussed in a broad and loose definition that ranges from communication of ideas and cooperation across disciplinary boundaries to different forms of knowledge integration in epistemological and methodological terms, integration of concepts and terminology, syntheses of data and knowledge, including education which is seen as an interdisciplinary practice (OECD 1972: 25). With this wide range of phenomena of interdisciplinary knowledge practices interdisciplinary knowledge practices develop in different, also competing, ways, including a variety of epistemic and organisational processes and forms of knowledge production and application for which rules and methods develop only slowly.

Interdisciplinary knowledge production developed outside academic research. It can be found already earlier in the history of sciences, although not under this name, for example, in political economy. Interdisciplinary knowledge production spread in the twentieth century, in several phases and forms, described by Jamison (2001) for environmental research and in the broader sense, in three phases of the organisation of knowledge production during the twentieth century: as "little science" (disciplinary research) before 1940, "big science" (multidisciplinary research) between 1940 and 1970, and since 1980 "technoscience" (including transdisciplinary research). Although this periodisation is not exact, it gives an idea about the changes in practices of scientific research that were insufficiently discussed in academic research, came with the unfolding of inter- and trans-disciplinarity. The discourse of transdisciplinarity, developing since the end of the twentieth century, deals with the integration of scientific and non-scientific knowledge, broadening knowledge integration and synthesis.

Environmental research shows the methodological difficulties of developing interdisciplinary research: it cannot develop without interdisciplinary knowledge syntheses, but syntheses across the boundaries of the social and natural sciences are epistemologically and methodologically more difficult than the limited interdisciplinary research within the social and natural sciences, across the boundaries of neighbouring disciplines that work often with similar methods and theories.

The epistemological characteristics of interdisciplinary environmental sciences can be described as follows:

Interdisciplinary research about society and nature developed during the twentieth century in several forms, where the newly developing discipline of ecology influenced the research, the conceptual frameworks and theories in the social and natural sciences. Interdisciplinary approaches include human ecology, cultural anthropology and cultural ecology, social and political ecology, developing as new interdisciplinary subjects with specific thematic profiles and specialisation, combining interdisciplinary knowledge production and specialisation. The more recent developments of sustainability science, resilience research and research on earth system governance bring new epistemological and methodological discussion into interdisciplinary research, especially with regard to complex systems as interaction social and ecological systems, for which the methodologies of research and forms of theory constriction in specialised disciplines are not applicable, as the following discussion of different forms of interdisciplinary knowledge production shows.

Interdisciplinary knowledge practices imply different forms and problems of knowledge synthesis that can be described more generally, with environmental research as a paradigmatic case:

1. *Consequences of specialisation.* The disciplinary fragmentation of knowledge through progressing specialisation makes knowledge integration and synthesis always weaker and limits synthesis to the field of specialisation and mainly to empirical research. With the progressing differentiation and specialisation of academic research knowledge syntheses were neglected and become ever more difficult, not only because of the rapidly increasing quantity of scientific knowledge, more because of newly developing fields of research where complex phenomena are studied. Complexity of problems, systems, and processes studied require interdisciplinary cooperation and knowledge synthesis, supporting the trend towards interdisciplinary research. Interdisciplinary research and knowledge synthesis grew with the experience that many problems investigated in specialised research can neither be explained nor solved through the limited and selective

knowledge created in specialised research. This general premise justifying interdisciplinary knowledge practices can be translated in the epistemological rule: not the discipline and not the general rules for scientific research determine the knowledge requirements for the explanation and solution of a specific problem, but the formulation of the problem determines the knowledge required and the rules for knowledge production and synthesis. The trend towards interdisciplinary science shows that the boundaries between disciplines are artificial, cannot deal with the "real life problems" as they are often called, that require broader knowledge syntheses.

2. *Transfer and exchange of data, methods, concepts, theories.* Such forms of knowledge exchange between neighbouring fields of research are already widespread in disciplinary research, showing a kind of "latent interdisciplinarity". The aggregation of data and knowledge from empirical research may be less complicated when data from methodologically similar studies are combined, for example, statistical data through meta-analysis. Knowledge syntheses with several, theoretical, empirical and normative knowledge components and syntheses of knowledge from natural- and social-scientific knowledge as in the research on anthropogenic climate change, become ever more complicated and are less structured through epistemological and methodological rules. A consequence is that complex knowledge syntheses are more dependent on the expertise, the cognitive interests, the individual assessments and the normative views of the researchers doing the syntheses, or on the consensus of cooperating scientists and other stakeholders involved.

The transfer of concepts from one discipline to another and the synthesis of knowledge created with these concepts is less complicated, although often controversial, as the example of the differentiation of the resilience concept in ecological, interdisciplinary and social-scientific variants shows or the interdisciplinary broadening of the concept of metabolism with that of societal metabolism (both discussed further in this book). More complicated is the development of theory for framing synthesised knowledge and creating explanations that require a combination of different theories from natural and social scientific research. In all forms of complex knowledge syntheses a main problems is the selectivity of knowledge used in the synthesis. In most disciplines competing paradigms, theories, and methods can be found; as a consequence of that exist competing explanations and interpretations of the problems analysed. The competing forms of knowledge are rarely discussed with regard to their potential integration; through their competition they appear in a power-perspective: which theories, methods and approaches are winning in the knowledge competition and advance to the accepted approaches in a discipline for a certain time, or are accepted in knowledge transfer and application? The selectivity of knowledge synthesis cannot be dealt with through logical rules and criteria alone, requires further procedures, rules and criteria that are hardly discussed epistemologically, only pragmatically, often are normative criteria used. A requirement for syntheses is then, to document the procedures and criteria applied, the decisions about theories, concepts and approaches applied in the synthesis, in similar forms as research protocols document the methodology of production and interpretation of the research results.

3. *The relations between specialisation of research and integration of knowledge.* Disciplinary specialisation and interdisciplinary knowledge integration are not necessarily contrasting developments, although the dominant view of interdisciplinarity is, that knowledge integration, synthesis, holism or unification of science are seen as countervailing the dominant trend of scientific differentiation and specialisation. In some interpretations, the phenomena discussed as interdisciplinary knowledge production are seen as specific forms and phases of the development of scientific knowledge within the dominant mode of disciplinary knowledge production (Dogan and Phare 1990). Although this is not developed to a systematic debate of the forms of interdisciplinary knowledge production in relation to the dominance if disciplinary knowledge cultures, it can help to connect disciplinary and interdisciplinary knowledge practices, showing that interdisciplinary practices encompass both phases of knowledge production, specialisation in the processes of analysis and synthesis in later phases: specialisation and synthesis are iterative phases of knowledge development in a broader view of the scientific knowledge process. With interdisciplinary knowledge practices integration and synthesis

are only broadened and given more importance as forms of knowledge generation, which is supported through the complexity of problems, systems, processes studied in interdisciplinary research.

4. *Interdisciplinary approaches in the social sciences* are described by (Klein 1990; 2007: 32ff.) as developing in a knowledge sphere where synthetic disciplines (as anthropology) and intersecting knowledge fields (as behavioural or environmental research) exist that are already transgressing disciplinary boundaries. Social-scientific interdisciplinarity developed in several phases after the First World War through new fields of research about problems that could not be dealt with in the limits of specialised research. This development happened through the synthesis of knowledge from two or more disciplines, thus creating new discourses. Interdisciplinary knowledge integration became part of research and knowledge production, whereas the teaching of science remained in disciplinary domains. This conventional view of interdisciplinarity was reinterpreted by Dogan (Dogan and Phare 1990; Dogan 1996): interdisciplinary knowledge production is seen as a component of the development of specialised disciplines and subdisciplines which advance in two phases of development, specialisation and reintegration of knowledge that includes interdisciplinary knowledge in the form of "hybrid specialisation". This interpretation differs from most forms of interdisciplinary practices. The discussion shows the varying conditions of knowledge production: the boundaries between disciplines in the social sciences are not as rigid and clear as in the natural sciences; many themes and phenomena are investigated in different disciplines and in parallel knowledge production. Theories are in the social sciences more important and more complex than in most natural scientific disciplines.

5. *Interdisciplinary approaches in the natural sciences.* General systems theory (Bertalanffy) developed in biology and spread later in discipline-specific variants of system concepts and system theories in many disciplines. It remained for long time in the twentieth century the dominant form of conceptual integration of knowledge from different disciplines through a common terminology and conceptual frameworks that allowed as well the description of social and ecological systems as more complex forms, as of simpler systems as

mechanical and cybernetic forms. Systems theory developed in the form of a scientific movement that established a terminology, way of thinking and research as a new scientific knowledge culture parallel to others. What makes the systems concept to a core concept of a specific form of interdisciplinary thinking and research is its high level of abstraction and generality that enables comparison between heterogeneous kinds of systems, for example between mechanical systems and ecosystems. In interdisciplinary research in ecology developed since the 1970s various interdisciplinary approaches that connect social- and natural scientific knowledge: human ecology, political ecology, social ecology, sustainability science, resilience research; these are the dominant forms of interdisciplinary knowledge production in environmental research.

6. *Transdisciplinarity* as a new approach began towards the end of the twentieth in the social sciences with a debate about new forms of knowledge production under the names of "mode 2" and "transdisciplinarity" that indicated an epistemic crisis in science (Nowotny et al. 2001). Also, this discourse developed from a scientific movement with normative views of knowledge production (Charter of Transdisciplinarity 1994, in: Nicolescu 2002), but adopted different forms in its further development. Transdisciplinarity is defined differently and in abstract forms. It is either seen as

- integrating diverse forms of research in many disciplines, in problem-oriented research and knowledge production, where not only scientists are involved, but other knowledge bearers, thus also integrating scientific and non-scientific forms of knowledge (Hirsch Hadorn et al. 2008); in this sense, it is connected with interdisciplinarity;
- or in a more normative view ("Charter of Transdisciplinarity") where it appears as different from interdisciplinarity through generating spaces for knowledge synthesis across, between and beyond disciplines. Transdisciplinarity creates new knowledge through broad syntheses, and it creates a new vision of nature and reality; in this view it is based on a value-loaded critique of specialised disciplinary science: disciplinary science makes a global view of the human impossible; it cannot confront the complexity of the world

and understand the spiritual and material self-destruction of the human species; it resulted in a technoscience that obeys the logic of efficacy for the sake of efficacy; it is engendering increasing inequality between the different nations on the planet (Nicolescu 2002: 147ff.).

Whereas these normative premises remain controversial, it is evident that they include in abstract form, a critique of science as it has developed in modernity and with modern capitalism, taking up motives from critiques of the "dehumanisation" or "barbarisation" of science. The manifold practices of transdisciplinary knowledge integration are not necessarily connected to this visionary thinking that makes transdisciplinarity into a new ideology of knowledge production, similar as consilience. From the ways of practicing transdisciplinarity it can, in spite of the radical critique of the disciplinary organisation of science, be concluded, that it is, as interdisciplinarity, another form of scientific knowledge production, but part of science. The common elements of transdisciplinarity as a new knowledge practice can be described as broad knowledge syntheses, problem-oriented knowledge generation and including in knowledge integration other forms than scientific knowledge, especially local and practical knowledge.

Whereas the different knowledge practices show that interdisciplinary knowledge production and synthesis need to be specified for concrete problems and thematic areas of knowledge, it is evident that most of them are lacking concrete knowledge-generating methods, instruments, tools. They cannot adopt the methods of empirical research in the manifold and broad forms of knowledge integration they aim at. Transgression of boundaries between the natural and social sciences creates problems of knowledge integration and combinations of different, scientific and other knowledge forms but pragmatic solutions have been found (described in Table 5.1, Chapter 5). The further discussion is focused to knowledge integration in environmental research and environmental governance, where a common theme is that of the interaction between modern society and nature.

6.2   Interdisciplinarity Across the Boundaries of Social and Natural Sciences in Environmental Research and Interdisciplinary Ecology

In the crossing of disciplinary boundaries and in interdisciplinary knowledge syntheses across the boundaries of natural and social sciences the question comes up, how to connect or integrate knowledge from ontologically, epistemologically and methodologically different sources and forms of disciplinary research that cannot work with the methods of aggregating data and empirical knowledge, but require combinations of different epistemologies and methodologies?

In Chapter 1, the differences between the natural and social sciences in diagnosing the global environmental and governance crisis have been described; natural scientific research was much less involved in this diagnose, although it is a main component of environmental research. It seems that there is a tacit division of labour between natural and social scientists in environmental research, as temporarily discussed for climate research: the first ones are responsible for the generation of new knowledge, the later ones for the transfer and application of knowledge in the practices of decision-making and governance. In the interdisciplinary knowledge practices in environmental research appear, however, further cognitive problems and problems of division of labour in knowledge production that reveal this view as unrealistic and simplified; it does not ask and reflect, how the roles and the forms of work of researchers and scientific knowledge bearers change when they become involved in processes of science communication, knowledge transfer and knowledge sharing for purposes of environmental governance.

The complexity of systems and processes to deal with in interdisciplinary research and environmental governance, support two doubtful methodologies of knowledge production: the inclination of simplifying complexity through modelling and the inclination to reduce the aspirations of knowledge integration and synthesis. An unsolved cognitive problem is, furthermore, how antinomies and contradictions are dealt with. The widespread practice is, to analyse contradictions logically and semantically and eliminating them from scientific knowledge, with the help of formal logics. This generalised rule system is based on the linguistic premises that are assumed to apply for all fields and forms of scientific knowledge production. Contradictions are accepted in paraconsistent logics (Bremer 1998) or in dialectical logics that exist as critical and minority discourses. Contrasts, incoherence, irreconcilable norms, interests and goals or conflicts are not contradictions, but they may result in contradicting forms of reasoning and behaviour, in clashing values and conflicts between actors. Governmental and political action is often contradicting because of the necessity to deal simultaneously with many, incompatible and non-reconcilable interests and norms. This is also the case with environmental policies that clash with other, especially economic policies.

In environmental research, with a large degree of interdisciplinary knowledge, a part of the contradictions are not recognised as such by the researchers; reasons for that are that the phenomena investigated are too complex and need to be simplified or modelled; that knowledge comes from different disciplines and fields of specialisation with different, discipline-, discourse- and theory-specific semantics. Knowledge formulated in different epistemologies, methodologies and semantics is not easily communicated across the boundaries of disciplines and specialised knowledge production. Interdisciplinary research is discussed with regard to the construction of joint languages and terminologies to communicate between researchers, researchers and knowledge users, and to transfer knowledge between disciplines. This is often done by way of simplification, without removing the communication barriers.

Although it is not developed for the purposes of dissolving inconsistencies in thinking and reasoning, conflict mitigation becomes an important procedure in natural resource management and environmental governance, where often such communication problems appear and need to be dealt with indirectly. It seems possible to develop conflict mediation in a more multifunctional perspective, where inter- and transdisciplinary communication problems are dealt with. The widely established procedures of mediation of environmental and resource use conflicts (Vasconcelos and Flávia 2015; Stepanova 2015) provide a basis for such "retooling", as also the research on conflicts in natural resource use with the instrument of reconciliation action plans (Klenke et al. 2013).

6.3   Variations of Knowledge Synthesis in Environmental Science and Social Ecology

Interdisciplinary knowledge syntheses in the environmental sciences include various forms of conceptual and procedural knowledge integration:

- connections of different theoretical concepts and theories in interdisciplinary environmental science, human, social and political ecology,
- the use of normative and value-loaded terms, including metaphors, in environmental science and social ecology,
- the connection of environmental research and governance with ethical reasoning,
- transdisciplinary knowledge integration along the knowledge chain— generation, dissemination, transfer and application of scientific knowledge in governance processes.

Describing knowledge syntheses generally, in epistemological and methodological terms that work for all problems and situations in form of "panaceas" is not possible. The knowledge synthesis problems are different and specific, therefore, a first step of developing a "toolbox" for knowledge integration is, to document and use experiences, case studies of successful synthesis projects, evaluation reports and epistemological reflections of interdisciplinary knowledge syntheses, thus also developing gradually methodologies and coming away from pragmatic and intuitive methods. The forms of knowledge synthesis described in the methodological literature are mainly for empirical research (Weed 2005) and for knowledge synthesis in disciplinary research (for ecology: Ford 2000). Syntheses in interdisciplinary environmental research vary along a continuum of weak and strong forms. Weak syntheses can methodologically be described as such, where the synthesis is not accumulation or based amalgamation of data (as, for example, in meta-analysis), but interpretation of different studies, research results, or concepts and their lose connection for a specific cognitive aim, resulting in combination of different perspectives that give together a better understanding of the problem of phenomenon analysed or close knowledge gaps. This is the case with many forms of qualitative synthesis (Weed 2005).

Strong syntheses are not necessarily aiming at a unification of science, universal explanations or theories, but broaden the scope of explanation and improve the understanding and explanation of complex problems or processes such as global environmental change that cannot be explained through a single theory, framework or method, require interdisciplinary knowledge and a series of epistemologically, theoretically, and methodologically different approaches to be understood (contextualised syntheses). Theoretical syntheses of knowledge from different theories are often limited to concept transfer, recombination of concepts or reinterpretations of concepts, without further integration of knowledge from specialised fields of research. Improving theoretical knowledge syntheses requires as a core component system analyses: theoretical analyses of societal systems (economy, politics), ecological systems (ecosystems and biomes at different spatial scales, from local systems up to the earth system), and of the specific forms of coupling between social and ecological systems (through practices of natural resource use and their economic transformation through the institutions of the modern world system).

In the synthesis of natural and social scientific knowledge in environmental research a series of problems come up that need to be discussed more specifically in the scientific knowledge practices, although they cannot always be dealt with in form of general rules and criteria:

1. *The status of concepts in interdisciplinary theories and explanations*: the concepts have different cognitive functions for research, theory construction and synthesis. Connections of different theoretical concepts and theories in interdisciplinary environmental science and social ecology require intensive discussion of questions like: How can theoretical concepts be transferred from the natural to the social sciences (so far the main form of concept transfer in environmental research)? How can explanations be created or improved through combinations of concepts and theories? How can criteria and rules

for problem-solving be derived from scientific knowledge and knowledge syntheses (including rules for transfer and application of scientific knowledge? How can rules and criteria for decision-making in environmental governance be derived from theoretical knowledge syntheses? For answering such questions "transfer projects" can become necessary—to provide opportunities for systematic reflection and experimenting with synthesis methods.

2. *The use of normative and value-loaded terms,* including metaphors, in environmental science and social ecology connects with the first question. Many new scientific concepts used in environmental research develop from metaphors usually transferring a concept from one theory or discipline to another, where it has not yet a clear cognitive role, but describes a phenomenon in form of comparison, similarity or analogy. The transformation of unclear metaphorical terms into scientific concepts is not only a logical process of linguistic transformations, implies the redefining concepts, attributing new meanings to them, specifying and operationalising them, defining methods for research and integrating new concepts into theories. The example of societal metabolism shows these processes in paradigmatic form. As long as it was understood as a metaphorical term, comparing economic or social systems with organisms, not much theoretical analysis was possible with this concept. Since it developed in social ecology into a new theoretical concept of socio-metabolic regimes, it has become a theoretical tool for analysing, comparing and explaining the socially organised forms of natural resource use in different historical societies. The concept of resilience did not develop in this way, brings more the epistemic and cognitive problems with interdisciplinary concept transfer from the natural to the social sciences, is stuck in the conceptual differentiation of heterogeneous forms of resilience.

3. *The use of ethical reasoning in environmental research and governance* implies forms of knowledge integration that cannot be seen as merging the different forms of empirical, theoretical and normative knowledge. Regarding ethics of knowledge use interdisciplinary knowledge practices struggle with many difficulties of combining empirical scientific knowledge and ethical reasoning, as the continuing debate about the relationship between facts and values in

science shows. In philosophy and ethics the forms of specialisation are different from empirical research where interdisciplinary teams of knowledge producers and users are found. The open, discursive and normative reasoning and discussion in ethics makes synthesis to a form of continued dialogue. Ethical questions in science have become important with the growing power incorporated in science, with research and technologies to modify and colonise nature as well as human nature. Similar knowledge problems come up with the ideas of transdisciplinary knowledge integration along the knowledge chain of generation, dissemination, transfer and application of scientific knowledge in governance processes, where knowledge is translated into different forms, not simply integrated.

Not much knowledge synthesis is done when the concepts of nature and society or social and ecological systems are connected, for example through hybrid concepts as "socionatures", or with the conceptual construction of social-ecological systems based on the assumption that both systems are integrated by necessity, as it is done in some forms of resilience research. It seems that with such combinations of concepts a lot of knowledge about society, social systems and their differences from ecosystems, is ignored, not controlling the knowledge production and application epistemologically and methodologically. Combinations of concepts are not yet knowledge integration, can at best be seen as heuristic rules for analysing the forms of coupling and interaction between ecological and social systems. With that knowledge integration comes into view, based on theoretically guided empirical research about the forms of interaction and the coupling of social and ecological systems. Similar forms of incomplete or speculative synthesis, at higher levels of abstraction, happened in the discussion and critique of conceptual dichotomies such as nature and society, or nature and culture (Goldman and Schurman 2000). The operations of knowledge synthesis are not done with model constructions and conceptual frameworks, but in methodologically guided integration, theoretical codification and limited forms of generalisation, for example, in historically and culturally specific forms and concepts as that of socio-metabolic regimes. When society and nature or social and ecological systems are conceptually integrated, it is necessary to specify the integration in terms of their coupling and connections. This is not done with the general assumption that hybrids are created in modern society, through human modification of nature. Also in the construction of hybrid concepts the differences between nature and society, social and ecological systems, living and nonliving systems remain visible, cannot be dissolved or levelled.

To exemplify these problems two examples are discussed in more detail: (1) the distinction between nature and society in the attempt by Ingold and Dickens to account for different views of nature and society in premodern and modern societies (Box 6.1), and (2) the theoretical synthesis of concepts of modern society in Giddens' theory of structuration (Box 6.2).

Box 6.1 Distinctions between nature and society in theoretical terms

Different views of nature can be found in the history of societies, and in the historical process emerges at some point in time emerged in the historical process a differentiation between the concepts of society and nature, attributed to the development of modern society where several epistemic and social ruptures could be observed. Ingold and Dickens refer to ethnographic research, differentiating inexactly between premodern and modern societies. Premodern societies are described by them as such, where humans usually did not distinguish between society and nature, only between other humans, non-human life of plants and animals, and inanimate nature. In modern societies humans usually distinguish between nature and society, creating a world of society where people live and one of nature where animals and plants live. In this person-centred view of nature and society humans are distinguished, rather separated as persons that are part of society and as humans or organisms that are like animals and plants part of nature. One can describe this as a further differentiation of language that creates in perception, thinking and speaking higher levels of abstraction, as one of the potentials of human language, as has been described in Chapter 2, with the consequence of "violence of abstraction". Yet, a linguistic explanation of this kind does not comprise all aspects of the distinction between society and nature, which includes not only linguistic and symbolic variants but differences in material reality.

The conceptual distinction between nature and society is not a sufficiently reflected and theoretically codified distinction that explains the development of interaction nature and society historically, in its specificities and changes throughout human history. It requires complementary concepts and differentiations at more specific and concrete levels of analysis, without simply dissolving dichotomies. The final explanation of the distinction between nature and society is one through the social processes of knowledge generation in two types of human-environment constructions: in premodern societies knowledge is gained in the lifeworld—processes of dwelling in the world of nature and interacting with it; in modern societies, knowledge is less gained through direct and personal interaction with nature, includes more abstract knowledge gained through "stepping out of the environment" (Dickens 1996: 4f, referring to Ingold).

This view and interpretation of human relations to nature shifts the discussion of knowledge generation from one of analysing nature–society interaction systematically in its material and symbolic relations to forms of normative and symbolic integration of humans and nature at the levels of views of nature and worldviews which characterise different societies and cultures. What would be required instead is to study the interactions between humans, society and nature in its varying material and symbolic forms historically, empirically and theoretically, for different civilisations and societies in human history, as it has only recently been done with the study of societal metabolism in social ecology. Then the distinction between nature and society is not only seen as one that appears only in the minds of humans as their views of nature and society, but one that is bound to their social practices of natural resource use, materialising in the societal metabolism which is not operating through single individual behaviour, but through systemic interaction of society and nature in such organised systems as states and governments, modes of production and economies. Humans make conceptual models about their relations to nature in form of abstract views of world and nature. These individual or collective views or constructions of nature should be differentiated from analyses of specific collective and material forms of natural resource use that characterise the societal relations with nature in a given society or for specific social groups.

Sources own discussion; sources mentioned in the text

The analysis of historically varying conceptualisations of nature, culture and society creates awareness of the multiple meanings of the terms. The sociological meaning of society develops fully within modern society only, deviating from older meanings of the term. *Modern society* is reflected theoretically in the social sciences in a variety of theories that developed for a long time outside of sociology (for example, in the political-economic theory of modern capitalism by Marx, or the presently influential world systems theory of Wallerstein), and in sociology after its establishment as a discipline further theories developed (see the overview in Bruckmeier 2016: 32ff.). A first sociological synthesis of theories was attempted by Parsons (1937) in a general theory of social action, later on in the critical theory of Habermas (1980), in the systems theory of Luhmann (1984), and in the theory of structuration by Giddens (1984). All of these examples are inter-theoretical syntheses of knowledge about modern society that try to integrate knowledge from different sources and theories. Giddens theory (Box 6.2) shows the epistemological difficulties of theoretical synthesis of theories of action and of systems in exemplary form.

Box 6.2 Interdisciplinary synthesis of theories of modern society—the example of Giddens

Giddens (1976) refers to the epistemological discourse about science in the twentieth century, where the relevance and the status of theory was discussed, with a minimal consensus between the theories of science of Popper, Lakatos, Kuhn and Feyerabend, supporting the conclusion that all forms of empirical research and observation are guided by theoretical assumptions, rejecting the idea of theory-independent research. As a consequence, Giddens formulates the first task of a social theory as that of developing the theoretical concepts that can frame empirical research. In the sociological discourse, core concepts are that of social action and social systems, that should be synthesised in Giddens' (1984) theoretical account of the constitution of modern society, including as well ideas from critical theory as from other theories of society, with reflections in epistemological and methodological forms. The contrasting concepts of structure and action or agency indicate the incompatible methodological perspectives of objectivism and subjectivism, resulting in theories of social structures (that exist independent from individual and subjective forms of action and perception) and of intentional social action. The relations between these two theoretical perspectives can be reconstructed differently, as the other synthetic theories in sociology show. Giddens addresses it in the form of a methodological dualism in which the contrast between both perspectives should be dissolved, connecting the concepts of structure and action without attributing primacy to each of them. The contrast between structural holism or objectivism (where theoretical explanation is generated in similar form as in natural-scientific research), and interpretation of social action (in the hermeneutic and subjectivist variants) dates back to the cognitive approaches in the classical sociologies of Durkheim and Weber.

Giddens attempts to synthesise both contrasting versions of sociological explanation by showing the generation of social structures through the social action of individual persons or subjects. This logical solution in a theory of structuration is a weak synthesis that leaves decisive questions open; the integration of objectivist and subjectivist, of macro- and micro-sociological perspectives is done at the price of modifying the concepts, causing further difficulties of explanation, the reflections oscillating between the perspectives. The theoretical synthesis cannot be done with the two concepts of structure and action alone, requires further theoretical concepts for creating sociological explanation, such as identity, agency, power, social system, social reproduction, time and space, reflexivity and reflexive modernisation, that need to be matched with the two concepts. The attempt to formulate a coherent theoretical logic of explanation results in modified and simplified concepts. The concept of structure is interpreted in less deterministic forms than in the theory of Parsons, where the actors appear as puppets on a string. The concept of system is reduced to a variant that can be formulated in terms of social action, as the production and reproduction of spatially and temporally interwoven actions. The concept of action is modified in its core component of intentional action, showing that finally intentions, the motivating and structuring component of social action cannot be reproduced within a systems perspective. The modifications of these theoretical concepts in Giddens theory show a simplification and abstraction that reduces the complexity of the systemic constitution of modern society from the sociological analysis to simple processes in which the abstract terms production and reproduction are hardly specified; rather than through synthesis the theory develops through linguistic manipulation of concepts and shortcuts in which the concepts lose much of their sociological meaning. This is visible in formulations as "duality of structure", a shortcut formulation of the genesis of society—social structures are created by human agency, the action of individuals that are at the same time limited by these structures—and in the concept of structuration. "To examine the structuration of a social system is to examine the modes whereby that system, through the application of generative rules and resources is produced and reproduced in social interaction" (Giddens 1977: 118). "Society only has form, and that form only has effects on people, insofar as structure is produced and reproduced in what people do".

To make the modifications and reductions of the concepts coherent, Giddens introduces normative assumptions and reasoning showing more his convictions and beliefs than a theoretical reasoning. He rejects functional and evolutionary theory as representing a non-historical reductionism and reification of structures, and also the historical materialism of Marx, in the last analysis opting for an interpretive and action-based approach where the shortcut message is society is produced through continuous social interaction, action representing the capacity of human actors as autonomous subjects.

Sources own discussion; sources mentioned in the text

An exemplary critique of the theories of Giddens by O'Boyle (2013) starts from the observation that many authors criticised Giddens for insufficiently elaborating in his methodological writings the relationship between agency and structure. O'Boyle wants to draw attention to another contradiction in Giddens work: whereas the early writings of the structuration theory tend to reduce structure to an epiphenomenon in celebrating the agent, Giddens later writings build around the diagnosis of the structural omnipotence of neoliberal capitalism, showing that this theorising oscillates between the alternatives of determinism and voluntarism that Giddens wants to avoid. Already earlier in the sociological debate similar critique has been formulated as an incoherence in the theory of structuration that tends to fall back in objectivist reasoning, cannot be integrated convincingly with the hermeneutic perspective of interpreting action (Kiessling 1986: 393). Whereas this and other critiques seem to see a successful integration of sociological theories and explanations as possible in future theorising, it can be doubted that such integration of action and system theory is useful or necessary. Giddens' shortcuts of theoretical terms and analyses reduced the capacity of explaining societal complexity.

It can be assumed that sociological theories and research perspectives require other forms of integration and synthesis than integration of abstract concepts. The theories of society represent different knowledge perspectives and methodologies for which integration does not necessarily result in a general theory. The theories explaining societal systems work with different perspectives, assumptions and interpretations of the systems, including also normative assumptions and premises. Different aspects of social reality cannot be amalgamated in one coherent theory and perspective; with regard to this the notion of general theory is misleading, an integration model of several complementary theories and perspectives creating further knowledge and explanation can be more adequate. An encompassing theory can be a combination of different theories (an inter-theory), with a plurality of concepts, perspectives, approaches and explanations that maintain their relative autonomy and explanatory capacity, which they would lose in a reductionist unity of theoretical reasoning. Open and plural forms of theoretical synthesis become important for complex and incompatible phenomena as nature and society.

6.4 Conclusions—Interdisciplinary Knowledge Integration and Differing Conceptualisations of Nature, Society and Nature–Society Relations

The theoretical integration and knowledge about nature and society, social and ecological systems, remains an unfinished and unfinishable synthesis, where the progress of synthesis and creation of new knowledge happens mainly through learning from the deficits and failures of prior attempts of theory construction.

1. *The interpretation of human relations with nature* includes many and changing variants in philosophical and scientific interpretations of nature and of the human conditions. What is passed down in the history of ideas, philosophy and science in forms of conceptualising nature and society are not all the cultural variations and views existing in former societies; these are only insufficiently known and selectively studied in ethnographic research. The historical knowledge is reflected, refracted and selected through the cognitive perspectives and interests of modern science, its epistemic rules, and the cognitive interests of specialised scientists. Also, the modern anthropocentrism in views of nature and society is a generalised construction influenced by the cognitive interests of scientists. Furthermore, different and contrasting conclusions can be drawn from anthropocentric thinking with regard to environmental consequences; it is not necessarily an explanation for environmentally destructive human behaviour.

2. *The main difficulty in synthesising knowledge about nature and society* is not one that can be dealt with in conceptual distinctions; it is given with the situation, that until now the largest part of knowledge about nature, society and their interrelations is existing only in specialised research in the natural and social sciences where the distinction between nature and society is not reflected in the research processes, but taken for granted, as in a lifeworld-perspective. As a consequence of that interdisciplinary knowledge syntheses across the boundaries of natural and social sciences struggle with different interpretations,

with incompatible epistemological and methodological forms of knowledge, and with a lack of rules for knowledge integration. To deal with these difficulties more systematically, not suffering from the "violence of abstraction", it seems meaningful to discuss the problems of theoretical synthesis separately for the two concepts of nature and society, and for the different methods of research connected with that in the natural and social sciences, only after that developing the further synthesis in theories of nature–society relations.

3. *Nature as reflected in the natural sciences* and their different disciplines as physics, biology, chemistry, geology, and many further subdisciplinary specialisations, is not dealt with in the form of a single, unified and synthetic theory. Whereas society can be described and explained in a coherent theory, the different concepts of nature indicate heterogeneous phenomena to be explained with this term, a plurality that cannot be reduced to one single and general theory. Nature, similar as culture, includes different components for which incompatible interpretations are created. The theories about nature comprise different aspects; in physics nature may as well be seen as the universe to be explained in astrophysical theory, as the earth, or as the microcosm of atoms; in biology the big theory of evolution comprises living nature; in ecology theoretical generalisations are limited; in geology, geography, anthropology compete for different explanations of nature. Philosophy has no longer the epistemic function to integrate all knowledge, but has become a discipline with manifold and competing theories and approaches.

Whereas attempts to conceptualise society in physical terms have rarely been convincing, attempts to understand and theorise society in evolutionary terms and forms exist since Darwin's biological and Spencer's sociological theory of evolution. All attempts to formulate overarching theories of nature and society remained controversial; at the end and as the only forms of theorising both spheres remain attempts to analyse and interpret their relations and interactions theoretically, not as part of general theories of nature, culture, humans or society, but in theories that explain the changing as relations and interactions between these spheres.

References

Bremer, M. (1998). *Wahre Widersprüche: Einführung in die parakonsistente Logik*. Baden Baden: Academia.

Bruckmeier, K. (2016). *Social-Ecological Transformation: Reconnecting Society and Nature*. Houndmills, UK: Palgrave Macmillan.

Dickens, P. (1996). *Reconstructing Nature: Alienation, Emancipation and the Division of Labour*. London and New York: Routledge.

Dogan, M. (1996). The Hybridization of Social Science Knowledge. *Library Trends, 45*(2), 296–314.

Dogan, M., & Phare, R. (1990). *Creative Marginality: Innovation at the Intersection of Social Sciences*. Boulder, CO: Westview Press.

Ford, E. D. (2000). *Scientitfic Method for Ecological Research*. Cambridge: Cambridge University Press.

Giddens, A. (1976). *New Rules of the Sociological Method: A Positive Critique of Interpretative Sociologies*. London: Hutchinson.

Giddens, A. (1977). *Studies in Social and Political Theory* (edition 2015). London and New York: Routledge.

Giddens, A. (1984). *The Constitution of Society: Outline of the Theory of Structuration*. Cambridge: Polity Press.

Goldman, M., & Schurman, R. (2000). Closing the "Great Divide": New Social Theory on Society and Nature. *Annual Review of Sociology, 26*, 563–584.

Habermas, J. (1980). *Theorie kommunikativen Handelns*. 2 Bände. Frankfurt: Suhrkamp.

Hirsch Hadorn, G., Hoffmann-Riem, H., Biber-Klemm, S., Grossenbacher-Mansuy, W., Joye, D., Pohl, C., et al. (Eds.). (2008). *Handbook of Transdisciplinary Research*. Berlin and Heidelberg: Springer.

Jamison, A. (2001). *The Making of Green Knowledge*. Cambridge: Cambridge University Press.

Kiessling, B. (1986). (without title, review of Giddens' theory). *Kölner Zeitschrift für Soziologie und Sozialpsychologie, 38*(2), 390–393.

Klein, J. T. (1990). *Interdisciplinarity: History, Theory, and Practice*. Detroit: Wayne State University.

Klein, J. T. (2007). Interdisciplinary Approaches in Social Science Research. In O. William & T. P. Stephen (Eds.), *The Sage Handbook of Social Science Methodology*. Los Angeles, London, New Delhi and Singapore: Sage.

Klenke, R. A., Ring, I., Kranz, A., Jepsen, N., Rauschmayer, F., Henle, K. (Eds.). (2013). *Human-Wildlife Conflicts in Europe: Fisheries and Fish-Eating Vertebrates as a Model Case*. Berlin and Heidelberg: Springer.

Luhmann, N. (1984). *Soziale Systeme. Grundriss einer allgemeinen Theorie*. Suhrkamp: Frankfurt.

Nicolescu, B. (2002). *Manifesto of Transdisciplinarity*. New York: State University of New York Press.

Nowotny, H., Scott, P., & Gibbons, M. (2001). *Re-Thinking Science, Knowledge and the Public in an Age of Uncertainty*. Cambridge: Polity Press.

O'Boyle, B. (2013). Reproducing the Social Structure: A Marxist Critique of Anthony Giddens' Structuration Methodology. *Cambridge Journal of Economics, 37*(5), 1019–1033.

OECD. (1972). *Interdisciplinary Problems of Teaching and Research in Universities*. Paris: OECD.

Parsons, T. (1937). *The Structure of Social Action*. New York: McGraw Hill.

Salmon, M. H., Earman, J., Glymour, C., Lennox, J. G., Machamer, P., McGuire, J. E., et al. (1999 (1992)). *Introduction to the Philosophy of Science*. Indianapolis, IN: Hacket Publishing Company.

Stepanova, O. (2015). Conflict Resolution in Coastal Management: Interdisciplinary Analyses of Resource Use Conflicts from the Swedish Coast. Ph.D. Thesis, School of Global Studies, University of Gothenburg.

Vasconcelos, L., & Silva, F. (Eds.). (2015). *Sustainability in the 21st Century: The Power of Dialogue*. Lisbon: MARGov-project.

Weed, M. (2005). 'Meta-Interpretation': A Method for the Interpretive Synthesis of Qualitative Research. *Forum Qualitative Research, 6*(1), art 37.

7

Social–Ecological Theory and Environmental Governance

To clarify the use of knowledge from an interdisciplinary theory of society–nature interaction for the practices and the improvement of global environmental governance it is necessary to analyse the relations between theoretical empirical, and practical or applied knowledge and the use of these knowledge forms in environmental policy and governance. The notion of an interdisciplinary theory is specified in the following discussion and the discourse of nature–society interaction is reviewed, as well as the distinction between nature and society in theoretical terms. The theoretical knowledge about modern society and its relations with nature, when it is used in the discourse of global environmental governance, should help to reduce the predominance of normative reasoning in the justification of global environmental governance and in the discussion of possibilities and pathways of a social–ecological transformation. The social–ecological transformation to sustainability requires visions of the future society and economy, but the normative ideas about sustainability, social and environmental fairness alone cannot guide the policy and governance processes. To begin the transformation it is necessary to understand hinders and difficulties of the process

© The Author(s) 2019
K. Bruckmeier, *Global Environmental Governance*,
https://doi.org/10.1007/978-3-319-98110-9_7

resulting from the "unsustainable constellation" of modern industrial and capitalist society.

7.1   The Development of Social–Ecological Theory

Theoretical knowledge about the present industrial society and the economic world system helps to identify the barriers of the transformation to sustainability. Ignoring such knowledge or relying only on empirical research about policy processes and failures is a major weakness of the present debate about environmental governance. Neither the empirical knowledge from policy analyses nor the empirical knowledge from ecological research provides information about the difficulties and the possible pathways of transformation. The theoretical analyses that create such knowledge are hardly discussed in empirical policy research that is oriented to the functioning or malfunctioning of specific policies, political institutions, or environmental regimes. The knowledge input from ecological research in policy and governance processes is limited to a few ideas and concepts as adaptive governance and "policies as experiments". The discourse of global environmental governance needs to be connected with that of economic globalisation and the possibilities of transformation of the economic world system. Much of the social-scientific knowledge that can be used in the governance process is from this research on globalisation and transformation of society and economy; it is this knowledge about the systemic constitution of modern society and its socio-metabolic regimes, and the knowledge about potential pathways of transformation that should inform the governance discourse.

As the beginning of attempts to make available knowledge about the systemic interaction between society and nature, although with imperfect methods and insufficient theoretical analysis, can be seen the Millennium Ecosystem Assessment in 2005: an assessment of the human impacts through natural resource use, modification of nature and pollution of the environment initiated by the United Nations. Since then the global scenario debate (Electris et al. 2009; Gerst et al. 2014) provided more and more specific knowledge about possibilities of the global transition to sustainability. With enriched knowledge and interdisciplinary knowledge syntheses it should be possible to discuss pathways of transformation more concretely than with normative criteria, worldviews and visions: with the help of theoretical and empirical knowledge provided from interdisciplinary environmental research.

That an interdisciplinary theory of nature–society interaction is relevant for the practices of environmental governance is not self-evident—and not widely accepted in environmental research and in environmental policy practices. Theory is usually seen as intra-scientific form of knowledge reflection with specific—but controversially discussed—functions such as description and explanation of empirical phenomena and processes, integration and generalisation of knowledge, or prognosis and identification of future development possibilities, trends and mega-trends. Especially with regard to the last, heuristic function of theory in the search of pathways to sustainability, there is much more theoretical knowledge available than used in the scenario-based assessments.

To specify the theory concept and separate it from that of model, theories are understood here as creating explanation through systematic analysis and interpretation of complex social and ecological phenomena, systems, processes; conceptual, graphical and mathematical models are mainly used to simplify complex systems, for simulating, calculating or projecting processes and changes, for production and application of knowledge. The differentiation between theories or theoretical knowledge and modelling is useful for describing more systematically the knowledge for environmental governance and combining the cognitive capacities of theories and models. So far global scenarios are the main method to generate transformative knowledge, blurring the differences between knowledge generated through theories, through empirical research, and through modelling. The concepts of theory and model are used in manifold ways; sometimes they are clearly differentiated; sometimes, in ecology too, theories are seen as forms of models, in conceptual or in mathematical forms. To differentiate between the forms of knowledge production in theories and modelling as suggested above, is sufficient for the following discussion.

The epistemic functions of explanation, knowledge synthesis and generalisation in an interdisciplinary theory of nature and society that uses knowledge from social- and natural-scientific research need to be specified for the different components of the theory, as described in Table 7.1. Such a theory has further purposes that are neglected in the theoretical discourse in the social and the natural sciences: the shaping and design of knowledge that is used in the practices of politics and governance. Only in the discourse of critical theory in the social sciences is—since its beginning in critical political economy in the nineteenth century—the

Table 7.1 Components of a socio–ecological theory of nature–society interaction

---

Structural analyses:

1. *Societal relations with nature;* for the analysis of socially organised forms of natural resource use, the abstract concept of societal relations with nature can be transformed in a conceptual framework for analysing natural resource use practices empirically and comparing them with regard to the environmental consequences (Bruckmeier 2013: 189). This historically comparative analyses is discussed in several examples in Chapter 1.

2. *Societal metabolism and socio-metabolic regimes:* The concept of societal metabolism has through the theory development in social ecology be transformed from a metaphorical term and thinking in analogies or through comparison of social systems and organisms, into a conceptual framework for analysing specific combination ns of natural resource use in different modes of production, especially for forms of energy used (Bruckmeier 2016). It is the core concept in social–ecological research that connects theoretical and empirical knowledge, to identify social and natural limits of human resource use and the transitions between different resource use regimes within and between historical and modern societies.

Process analyses:

3. *Colonisation of nature,* human modification of nature, and the forms of historically different natural and ecosystems produced through human work and technologies (Bruckmeier 2013: 221f). The presently relevant research includes the macroscopic research on global environmental change and microscopic research on genetic modification.

4. *Societal transformation (society and economy) and social-ecological transformation:* the presently intensifying discourse about transitions to sustainability through social–ecological transformation of modern society broaden the social–ecological discourse, connecting it with further research on transformation (in political ecology, ecological economics, political economy, research in sustainability science and on global environmental governance.

Sources Bruckmeier (2013, 2016)

---

necessity of theory for social practice and action constantly reflected. The theory appears, in the last analysis, not for scientific purposes of structuring and systematising knowledge, but for social practices, possible forms of social intervention, regulation, or transformation, to change the modern society and its economy, also through supporting social movements in the processes of social emancipation. This critical discussion emerges again in the discourse of transformative science for purposes of sustainable development, and the controversies about this idea resemble the prior controversies between traditional and critical theory in the social sciences (see the debate in the Journal "GAIA": von Wissel 2015; Rohe 2015). The distinction between scientific and practical knowledge in the natural as well as in the social sciences is often simplified and reduced to science–policy relations, assuming that the practice of social change and transformation is that of politics. With this inexact assumption the forms of practice are blurred and reduced to a specific form, that of policy or governance. The different knowledge practices in science and politics are not further clarified, the translation and transformation of scientific knowledge for its application and practical use are reduced to knowledge transfer and science communication (see Chapter 5). However, in the translation and transfer processes only limited parts of theoretical knowledge are applied, and this knowledge is often simplified. The differences between the two paradigmatic views of science and practice in critical and traditional theories are blurred in the recent development of environmental research, especially with the discourse of interdisciplinarity. It can be expected that from this interdisciplinary knowledge culture develop new forms of combining theories of different kinds, traditional and critical theories of society and of nature, to improve the knowledge for environmental governance.

Departing from these reflections about theory can be discussed the potential functions of an interdisciplinary theory of society–nature interaction for the practices and the improvement of global environmental governance. For this purpose, it is necessary to specify the notion of interdisciplinary theory and to review the discourse of nature–society interaction, as well as the distinction between these concepts of society and nature in theoretical forms. One purpose of discussing the use of theoretical knowledge about nature and society for the practice of governance is, to reduce the predominance of normative reasoning in the justification of global environmental governance. Such reasoning does not just provide ethical justifications, more often come with these justifications misleading and doubtful assumptions about the necessity of global environmental governance and its forms; theoretical and empirical knowledge seems to provide a better knowledge basis for governance than worldviews, visions, or normative criteria which are necessary but not sufficient. The term of sustainable development is open for interpretation with the help of theories, not only for normative interpretations prevailing in the practice of policies; also policies should be based on scientific, empirical and theoretical knowledge. The concept of environmental justice, in difference to sustainable development, highlights a normative dimension of environmental policy and governance, and directs the knowledge use more away from social theories, towards philosophy and ethics.

The variant of a theory discussed here is not the only one that can be used for the purpose of discussing the relations between science and environmental policy or governance. Yet, it is one that refers to and connects several theories. This theory, described in its development elsewhere in more detail (Bruckmeier 2016), is specific as a composite theory:

- it is an interdisciplinary theory constructed with the use of other concepts and theories, mainly from sociology, political economy, human and social ecology;
- it is a historically specified and differentiated theory, with different components, levels of analysis and forms of generalisation and explanation;
- it is not a finished theory, rather one in continuous development and modification with the integration of new knowledge.

Such a synthetic theory uses different components of theory construction: knowledge about social and ecological systems and their interaction; epistemological models of theory that can be described in preliminary terms as inter-theory, multi-scale theory, and theory of complex systems; methods of system analysis and assessment as they are practised in social- and in natural-scientific environmental research.

As a historically specified and situated theory this theory of nature–society interaction differs from conventional theories of society in sociology and conventional ecological theories through its focus on relations, interactions, interchange and interfaces between social and ecological systems. Such a theory is not aiming to replace other theories from different disciplines relevant for social and environmental research, but complementing them through more systematic analysis and explanation of the inter-systemic relations between natural and societal systems. Although it is a critical theory that is not compatible with all others in sociology and ecology, it uses more knowledge from different theories than others dealing with questions of nature and society. Main influences are from the following theoretical discourses:

- with regard to social-scientific theorising about society it can be seen as influenced from the broad tradition of critical theory in the sense specified by Horkheimer (1937) to which, beyond the Frankfurt School, can be counted more recent variants of world system analysis and theory by Wallerstein, the variants of ecological Marxism (see discussion in Chapters 2 and 3), the theory of the capitalist world ecology (Moore 2015), the critical theory of globalisation (Sassen), and the presently developing theoretical discourse about social–ecological transformation (Brand and Wissen 2017; Görg et al. 2017);
- with regard to natural-scientific and interdisciplinary theorising about ecosystems and their interaction with social systems, the synthetic theory of evolution from biology is less relevant; it has the function of a "background" theory clarifying the differing forms and temporal scales of societal evolution, biological evolution and the evolution of the earth system. Important theoretical discourses include that of complex adaptive systems, of co-evolution of society and nature (Norgaard 1988), of systems ecology, of human-ecological theories of human adaptation (Moran 2006), the emerging theory of the Anthropocene, and the interdisciplinary discourses of sustainability science, resilience research, and research about common pool resources (Ostrom et al.).

The social–ecological theory of nature–society interaction that emerges from this intertheoretical knowledge synthesis can be understood in its functions as a framing and knowledge integrating theory. Its main purpose is to systematise and integrate the dispersed theoretical knowledge about the relations between modern society and nature that is generated in different, disconnected discourses in philosophy, sociology (especially environmental sociology), political economy, ecological economics, human ecology, cultural anthropology, political ecology and social ecology.

The core component of the theory with the concepts of nature–society interaction and colonisation of nature is building on historical analyses of the relations between society and nature in different civilisations and societies in human history. These core concepts are specified further in operational concepts and conceptual frameworks as that of socio-metabolic regimes, including energy regimes, economic regimes (accumulation regimes), and international political regimes in environmental governance. The components of this theory are summarised in Table 7.1.

7.2   A Brief Review of the History of Changing Societal Relations with Nature

In Chapter 1 the history of human modifications of nature has been described which appears, in theoretically more elaborate form, as changing "societal relations with nature". This formulation, rooted in the philosophical discourse that is one of the sources of critical theory (Hegel and the early writings of Marx), developed with critical theory in the nineteenth century and continued to be used in the broadening discourse of critical theory in the last century. The theme of society–nature relations marked the contours of a theory of society claiming to clarify the systemic relations between nature and society, which cannot be analysed within the limits of disciplinary specialisation and in the newly emerging academic discipline of sociology. The basic argument of the theory is: there is no society without nature, and the form of society and its economic mode of production cannot be explained without a theoretical analysis of the interaction of society and nature through human use of natural resources in socially organised and changing forms. To understand the modern capitalist society requires not only knowledge about social structures and processes within society, but knowledge about the processes in which society and its members interact with nature through the use of natural resources, in material production in agriculture, industry, and other forms of natural resource use.

In the analysis of historical modes of production or socio-metabolic regimes in Chapter 1, a long trajectory and directionality of change has been identified that is maintained through all societal transformations in human history. The trajectory can be described as intensifying of natural resource use and the societal dynamics in the "*longue durée*" of history as interchange of long phases with evolutionary societal development and shorter transformational phases cumulating in "Promethean revolutions". Natural resource use and interaction between society and nature appears in this ecological perspective as growing in complexity, in terms of intensification of natural resource use and intertwining of society and nature—contrary to popular views in modernity, that society has, after many thousands of years of struggling with the forces of nature, successfully emancipated from nature and its determining forces, one more time revitalised with the theory of network society of Castells. Even the emerging theory of the Anthropocene is ambivalent in characterising the human relations with nature as dissolving or integrating society and nature. Whereas in the conventional form of thinking about nature in modern society the conclusion is that of an emancipation of humankind from the forces of nature and an autonomy of modern society, in the critical variant of the Anthropocene theory the detachment from nature is seen as a historical fact, but explained as maladaptive ecological change and requiring the reconnecting of society and nature and changing the social organisation of natural resource use towards sustainability. Both arguments, the autonomy and the dependency-hypothesis, seem to have as common weakness the abstract reasoning in terms of humankind as a collective subject, more visible in the theory of the Anthropocene. This strategic abstraction from the specific forms of organisation of societies directs the reasoning away from the identification of historically specific social subjects of change and transformation, from the societal division of labour and differentiation of social classes, or the development of social systems and institutions into "quasi-subjects", especially capital or the markets in the modern world system that control and direct the human agents.

The abstractions can also be seen as struggling with the complexity of societal relations with nature that are coded in historically specific socio-metabolic regimes. As with many other scientific terms there is no consensus about the term of complexity beyond the simple observation that systems described as complex include many parts that interact with each other in multiple forms. The alternating phases of slow evolutionary change and fast transformative change seem to reach another quality in modern society, where, in a very short historical and evolutionary time of several hundred years, processes of social and environmental change accelerated in hitherto unprecedented degrees, modifying nature and causing global environmental change in dimensions that have not been possible before, in hundreds of thousands of years of human history. The differentiation between slow evolutionary and fast transformative changes that characterised earlier societies and their interaction with nature can no longer be easily made; rather it can be said, the transformative dynamic of development has become permanent. Accelerated development and human modification of nature create now accelerating changes in nature, paradigmatically climate change. Some of the effects can already be perceived and experienced, especially in the increase of extreme weather situations. Other effects like sea-level rise and melting of the polar ice may show their consequences only after many hundreds and more years.

In the course of human history different societal relations with nature developed that can be specified and classified in theoretical terms of the historical modes of production, socio-metabolic or socio-ecological regimes. In social ecology the metabolic regimes are analysed for three historically important modes of production that structured the long historical epochs of human societies—the epoch of hunters and gatherers, of agricultural civilisations, and of modern industrial society. Further subdivisions and specifications of metabolic regimes are possible. Using the overarching term of societal relations with nature, metabolic regimes can be described for different social groups and forms of production that are still found in modern society which is characterised through the coexistence of a variety of modes of production that developed in earlier phases of society. The peasant mode of production, a form of simple commodity production (Chayanov), was existing for a long time in human history, in different cultures and countries, survived also in the Western industrial countries where it was gradually subsumed in the overarching capitalist mode of production. It was integrated into industrial capitalist production especially with the Fordist accumulation regime that spread after the Second World War characterised through mass production, mass consumption, and use of fossil energy sources.

When something important can be learned for environmental governance from the analysis of prior societies and their socio-metabolic regimes, it is not the specificity or temporally limited existence of all societies and modes of production. More important seem three connected points:

1. The duration of socio-metabolic regimes and their possible continuation in future (their "safe future", before they reached limits of global resource use) show significant differences between the historical societies: earlier societies had a longer "safe future" in terms of reaching the global limits of natural resources in the earth system. The industrial society has already "used" its future and is, after two hundred years transgressing the global boundaries of resource use.
2. The historical analysis and comparison of modern and earlier societies and socio-metabolic regimes helps to understand potential pathways to a future sustainable society. The comparison with the past enables thinking about possible changes, to identify new forms of transformation and transition.
3. The comparison with earlier societies helps to understand the limits and the traps of technological innovations: rarely were new technologies and technical inventions environmentally sustainable. Most new

technologies helped to solve older problems and to overcome temporarily limits of resource use, but they created new and unforeseen environmental damages.

Nature–society interaction in modern capitalist society. Since the neoliberal deregulation and globalisation began, simultaneous with the first global resource crisis (oil crisis in the 1970s), intensify also the environmental conflicts through the societal relations with nature inscribed in the capitalist mode of production and its metabolic regimes. The culturally and economically specific relations with nature in terms of worldviews and lifestyles of specific social groups and cultures vary more than the systemic relations with nature structured in the mode of production. Still larger parts of the population in countries of the global south, not yet minorities, live in pre-industrial modes of production as peasants and smallholders, artisanal producers, through small-scale local fisheries, as also ethnic groups and first nations; such forms of subsistence production and livelihoods, investigated in cultural anthropology, have "no future". Levi-Strauss who studied such cultures intensively came to the desperate conclusion, they are doomed to vanish, are destroyed with the spreading of the capitalist mode of production that has now reached all parts of the world, is valorising the last natural resources that were not yet subsumed under its metabolic regime—the tropical rains forests and its genetic resources, the deep sea, and the polar regions.

This doomsday prophecy, echoed in the ecological discourse about the environmental and self-destruction of the industrial society supports the ecological critique of growth: economic growth undermines the natural resource base and the conditions of survival of humankind. Justified as such critique is, ethically and in terms of scientific knowledge, it is not adequate and sufficient to develop the ecological discourse and motivate the projects of sustainable development and social–ecological transformation. It is not sufficient to understand the necessity of transformation, more important is, to find ways to initiate it. At this point begins the discussion of the cognitive functions and the practical use of an interdisciplinary theory of the relations between modern society and nature.

7.3 Enlargement of Knowledge for Environmental Governance

The development of an interdisciplinary theory of nature–society inter-action discussed in this chapter creates emergent cognitive properties in two forms that cannot be created by the more limited sociological and ecological theories from which it is synthesised:

- it helps to explain the complex interactions of social and ecological systems in modern society and the possibilities of their co-evolution in the long-term future, through a comparison with earlier forms of society and their societal metabolism, and
- it helps to explore potential pathways of social–ecological transforma-tion of modern society, a theme which is widely neglected bot in the theoretical discourses about society and nature; for the discussion of the future no scientific knowledge seems available; the future of mod-ern society is only discussed in some specialised fields of research as future studies and with regard to sustainability in the global scenario debate.

Sustainable development and global environmental governance are the knowledge practices and forms of social action, where knowledge about the future is required, where questions of coping with uncer-tainty and risks, with potential future events that are influenced from present social practices and practices of natural resource use need to be discussed. In the social and political practices of governance the future becomes a question of decisions to make or to avoid. To clarify the possible forms of dealing with the future in attempts to influence and redirect the development of modern society towards more sustainable pathways, and the significance of theoretical knowledge for this pur-pose, it seems useful, to identify the knowledge deficits and neglected themes and questions in the scientific and political discussions of social transformation, the future of society, and pathways towards sustaina-bility. Knowledge about trends of social and economic development and megatrends (global social and environmental change: economic globalisation, urbanisation, land use change, climate change, biodiversity reduction) that influence these trends is the basis of collective action and decision-making in policies of sustainable development and in global environmental governance. This knowledge (discussed in the preceding chapters) is not sufficient for the discussion of long-term trajectories of development and transformation. The knowledge allows only for planning and designing the immediate future as consequence of present decisions. Furthermore, it can show the behavioural, institutional, technological, infrastructural and systemic lock-ins, the dependency on growth-based economic development as the development path of the economic world system:

- the economic lock-ins in the capital- and technology-based forms of economic production and resource use;
- the lock-in in a material- and energy-intensive pathway of resource use and its technological infrastructures (power plants, large-scale technologies, large technical grids and networks for distribution of energetic and material resources);
- the lock-in in the industrial energy system on the basis of finite and fossil resources;
- the behavioural lock-ins in resource-intensive forms of private and public or collective consumption;
- the lock-in in urban development (mega-cities) and its dependence on long-distance supply systems.

The lock-ins that block sustainable development can also be reformulated as the forms of change and transformation that seem impossible in the capitalist economic system:

- the impossibility of degrowth and a transformation of the economic world system in a non-growing economy, described in older political economy as "stationary state" and renewed in ecological economics (Daly);
- the impossibility of a system transformation of the economic system and the supporting political systems, for which the lacking success of sustainability policies up to now provides a non-intended proof;

in the more ideologically than ecologically substantiated ideas sustainability is impossible because of the complexity of the systems in nature and society, not because of the unsustainable organisation of the modern economic world system and its world ecology.

These lock-in situations, obstacles, and development dilemmas are also the widely ignored problems in the public and political discourses of sustainable development and global environmental governance. To bring these questions and problems in the sustainability and the environmental governance discourses seems unavoidable, although difficult, because of practical reasons:

- Many political and economic actors and organisations, governmental institutions, many citizens and a part of the scientists and of the environmental and social movements do not see the necessity or the possibility of a transformation of the economic system. They accept the difficulties of environmental action and governance as a reality taken for granted; the lack of scientific knowledge or the complexity of natural and social systems in environmental sociology (Gross and Heinrichs 2010) and ecology provides arguments for the impossibility of system transformation. Furthermore, the lack of practical experience for sustainable development is often used in arguments for reducing the aspirations, accepting the limited possibilities, or referring to the necessity of experience and knowledge that needs to be gained in future, in form of "trial and error", "policies as experiments", "muddling through", indirectly arguing with the "veil of ignorance" or the unknown future. The limited interest of parts of environmental movements, but also natural scientists and political decision-makers in in social-scientific and theoretical knowledge, shows an unwillingness to deal with the complicated cognitive problems. Idealism and naivety of certain forms of environmental thinking and action have similar consequences of ignoring the difficulties of solving environmental problems. Abstract and vague ideas for sustainable development used in environmental politics add up to the difficulties; the terms are used with different interpretations and intentions, and the ideas are misused (Adams 1990: 53, with regard to the earlier strategy of eco-development; Bruckmeier 1994: 167f., with a broader perspective of global environmental policy).

- The vested interests of global players and powerful political actors who profit from the economic system and its social and environmental dysfunctionality, block economic and social–ecological transformation. This can go so far as to gain from environmental destruction in economic and monetary terms through the restoration of ecosystems and technical clean-up of pollution, transforming also environmental destruction into possibilities of economic growth and development.

- The asymmetrical political and economic power relations in the modern economic world system, widely known that and easily understood, without complicated theoretical analyses and reasoning, provide pragmatic, sceptical and disappointed reactions with forms of reasoning of the kind, it is impossible or useless to change something (sceptical environmentalism).

To come out of the dilemmas and traps and to deal with the inconsequent reasoning in environmental politics and governance arguments are required for the possibilities to break vicious circles of thinking and reasoning in the policy discourses. This is the main question in the further sustainability discourse that has so far hardly dealt with the dilemmas, or only in abstract forms. The first arguments are answers to the question, why interdisciplinary knowledge synthesis and critical theories like the social–ecological theory of nature–society interaction should help to improve global environmental governance and open chances for a new "great transformation"? Answering this question would also improve the possibilities to rethink other forms of knowledge use and knowledge combination in environmental governance.

The arguments for the utility of the systems analyses of societal and ecological systems (as part of the theory of nature–society interaction) for the improvement of strategies of global environmental governance are fourfold:

1. the system analyses (of ecological systems with regard to the progressing environmental destruction, and of societal systems, identifying

the reasons of environmental destruction through natural resource use) create knowledge of and make aware about the historical reality of the society and the state of the systems that need to be transformed on the way towards sustainability. This is first and foremost the economic world system and its connected systems of natural resource use, the political systems with national states and international regimes.

2. The system analyses show, how the dependency-relations between the global north and the global south, and the relations between national economies in international trade and exchange are generated and reproduced through the economic world system (valorisation and commercialisation of natural resources, unequal economic and ecological exchange).

3. The system analyses show, how nature and ecosystems are modified and transformed through the socially structured processes of natural resource use, with the consequences of degradation of ecosystems and global environmental change.

4. The system analyses show the interaction between the economic world system and the political systems, national states and international regimes, the political shaping of processes of societal and economic development—how processes of natural resource use are structured politically and in terms of legal rights (through property rights, environmental and other policies). Much of this knowledge cannot be generated through policy analysis and evaluation of policies. Furthermore, this knowledge shows that the normative reasoning, worldviews and visions, and the abstract concepts which prevail in the political discourse about the environment ignore, distort and misinterpret ideologically the systemic structuring of nature–society relations.

In Chapter 4, the crisis of environmental governance and the limits of environmental agency were analysed, including empirical research and knowledge about environmental agreements and regimes. This preparatory analysis of environmental governance can be systematised and structured with the help of the theoretical knowledge from system analyses. Less the empirical knowledge from policy research, more this knowledge from critical theories and systems analyses provides a solid scientific knowledge basis for global environmental governance and global transition to sustainability. The present debate about social–ecological transformation develops and specifies the knowledge from the system analyses further: there is already an intensive discussion about how to initiate social–ecological transformation and which knowledge is required in the policy and governance processes.

The empirical policy research on effectiveness of policies and international regimes is not devalued through the theoretical systems analysis, only specified and limited in its validity: it is adequate for showing the development in specific policy fields through monitoring and evaluation, but does not know about the contexts and systemic conditions under which environmental governance happens; the policy analysis cannot replace more theoretical analyses of nature–society interaction. Based on the arguments for a theoretical system analysis, it is possible to develop and specify the use of theoretical knowledge for global environmental policy and governance further (see: Bruckmeier 1994: 76ff.; 2013: 235ff.; 2016: 385ff.).

1. The system analysis in the social–ecological theory of nature–society interaction provides the arguments why the solution of global environmental problems and reduction of negative environmental and social impacts of global environmental change is not possible without a social–ecological transformation of modern society and the modern economic world system; concrete reasons are given in the description above of the lock-ins and maladaptive ecological change through a growth-based economy with the consequences of depletion of the natural resource base from which the future of the global society depends. Assuming, that a further expansion of the growth–based economy through the colonisation of the cosmic space and other planets is not a realistic option in this century, all problems need to be solved on this planet, through the sustainable use and management of natural resources.

2. Global environmental governance is with this broadening and contextualisation of the knowledge about the state of societal and ecological systems confronted with further problems which cannot be

ignored in environmental policy; it is connected with the transformation of the economic dependency-relations between the countries in the global north and south. The social–ecological transformation is simultaneously an economic transformation which is in conflict with the economic globalisation process. The north–south divide and the dependency relations did not vanish through economic globalisation, and new inequalities as the digital divide came. It becomes an additional task of global environmental governance with its framing in the sustainability discourse and strategies, to deal with the ideas of inter-generational solidarity, to support the achievement of the eradication of poverty, a millennium development goal of the UN. This goal refers indirectly to the global north–south divide, however not in a way that allows the combatting poverty effectively, more in the conventional forms of development aid. Without a theoretical system analysis as discussed here it will not be possible to understand that poverty abatement requires the transformation of the economic world system; it cannot be achieved in the same system that generates poverty as a requirement of its functioning, and it cannot be achieved with the normative idea of intergenerational solidarity alone, that gives only a vague idea that this requires global redistribution and sharing of resources. Seeking for ideas how to support politically the achievement of this goal, it seems necessary to take up the discussion of the reform of the international institutions including the UN which is also discussed in the discourse of Earth system governance. Already a long time ago, Altvater (1991: 366) demanded a broadening of the participation in environmental politics beyond the informal participation of different stakeholder groups: legally formulated participation rights of the people in all countries, to reduce the dependency of the international system from the governments. This is not a path-breaking idea, but at least one that brings the debate about stakeholder participation away from the informal and weak forms as it is practised now.

3. The global future in terms of inter-generational solidarity or sharing the natural resources on the earth with future generations, in an indefinite future, is a vague idea from the sustainability discourse, which can be specified with the knowledge from systems analyses

and from scenario analyses. It needs to begin with intra-generational solidarity of resource use, sharing and redistribution of natural resources between the presently living generations, including global redistribution between the countries and the global north and south. Redistribution and sharing of resources cannot be successfully organised through policy and governance systems; they require the development of new economic institutions beyond the market that is the dominant institution now. This debate is also of theoretical nature, and it is weakly developed in economics. The most important ideas are still the ones from Polanyi who developed also the concept of the "great transformation"; in the context of institutional economics he discussed the possibilities to develop new forms of reciprocal exchange and redistribution as compensatory institutional mechanisms to the dis-embedded markets. The discussion is taken up in the environmental discourse more than in economics.

4. Corresponding with the ideas of intergenerational solidarity and poverty abatement is that of broadening the concept of global commons as a means to reduce resource depletion and environmental disruption. Global commons would not include only the oceans, the atmosphere and stratosphere, also the arctic and antarctic ecosystems and the biosphere. The discourse about the commons and common pool resources has gained intensity with the environmental discourse in the past decades and is now an established field of interdisciplinary research (Ostrom et al.). Also this knowledge could be used more systematically in the discussion of global environmental governance. The critical argument from this research is, that not all natural resources can, as suggested in Hardin's (1968) analysis of the "tragedy of the commons" be transformed into private or state property as an effective means to prevent environmental destruction: also these forms of property cannot prevent effectively overuse of resources and environmental destruction, they are part of growth-based economic systems. The strengthening of the collective property form of the commons and of the responsibility of all resource users can support more environmentally sustainable forms of resource use and management.

5. The reasoning with ecological knowledge i.e. part of the points mentioned above is not efficient, when ecological, biological and physical "laws of nature" are seen as eternal and universally valid laws that need just to be obeyed in human resource use (Bruckmeier 1994: 270). These laws do not speak for themselves and they cannot simply be followed by resource users in the daily routines of natural resource use; the division of labour is so advanced that the "responsible users" creating the environmental damages cannot be easily identified, as the resource use processes are institutionalised and part of the political and economic systems. Ideas to use biological knowledge directly, in fisheries management for example, are only of limited utility, cannot deal with the economic incentives to overuse the resource and the ecological damages to ecosystems through pollution. In all such concrete and limited approaches of resource management with selective knowledge use there are more complex mechanisms and processes in social and ecological systems to take into account. In the final analysis, local strategies of natural resource use and management need to be set in the context of the societal and ecological systems and to deal with the complexity of the systems, if the aim of sustainble resource management should be achieved.

Thinking the future of environmental governance theoretically. The theories of society and of society–nature interaction end with the diagnosis and description of the present state of society and the environment; they leave the thinking about possibilities and ideas of transformations of society and of coupled social and ecological systems to other forms of cognition, thinking, knowledge production. Transformative governance as it is discussed within the global scenario debate, implies sophisticated forms of temporal structuring and combinations of different temporal structures in an overall picture of long-term transformation processes:

- the periodical and cyclical processes of ecosystem changes;
- the linear temporality of the development of social and economic systems in the short-term perspectives of growth;
- subjective perceptions of time in specific sociocultural groups; and

- the "*longue durée*"-perspective of historically long processes of evolution and transformation of societies, which offers possibilities to connect the societal development with that of changes in nature or the earth system.

These different temporal perspectives do not come together in one overarching, complex temporal perspective that guides the development of complex interacting systems. Theoretically conceived long-term transformations to sustainability, in several phases, appear as changing combinations of different time regimes; an exclusive time regime of sustainable development does not exist, only an overarching or framing perspective of a process that can take several hundred years, and needs to be broken down indistinct phase where the temporality can be structured for limited times. Sustainable development remains a combination of different time regimes, of cyclic, repetitive, linear and nonlinear changes, ruptures and revolutions, reversible and irreversible processes. As a consequence, governance strategies need to construct and combine differentiated, multi-scalar temporal and spatial regimes, which differ according to the knowledge used.

Differentiating the temporal perspectives in action for the future in global environmental governance is not sufficient in the inexact form of short-, mid- and long-term perspectives and timeframes that remind of the forms of planning for the near future. A review of different approaches and methods of constructing time and accounting for the future, discussed in social and environmental research, is required (Hassard 1990; Hassan 2009; Rosa and Scheuerman 2009). The broad debates of thinking about and acting for the future are methodologically reflected in future research and scenario analyses (Electris et al. 2009; Gerst et al. 2014).

The different methods of future studies (prognoses and forecasts, trend extrapolation, simulation, modelling, envisioning, scenarios) are not all relevant for the discussion of social–ecological transformation and global environmental governance. The most importants remain the global scenarios that developed since the Millennium Ecosystem Assessment in 2005. The scenarios can be methodologically improved through connections with further knowledge, especially knowledge about planetary boundaries of resource use (as has already been done: Gerst et al. 2014), and theoretical knowledge about society–nature interaction and social–ecological transformation (that is the main debate in the near future).

7.4 Conclusions—Reconnecting Society and Nature Through Environmental Governance

The ideas to broaden knowledge use in environmental governance are not simply an argument to put ever more knowledge and information in the governance systems, which would be a way to let the systems collapse through information overload, competing and contradicting forms of knowledge to be applied. The complexity of governance in the social–ecological transformations to sustainability needs to be reconstructed in different knowledge-related procedures, beyond the conventional forms of research and knowledge transfer or science communication. The rationale of including theoretical knowledge from system analyses in the governance procedures is twofold:

(a) reducing the normative reasoning and visionary ideas in the ecological and sustainability discourses where they have gained strong influence, often replacing or reducing the use of positive scientific knowledge, or replacing theoretical arguments; the knowledge use should be balanced between theoretical, empirical and normative knowledge components with clear functions in the governance process;

(b) differentiating the forms of knowledge production, knowledge communication, application and sharing, and recombining them in effective forms in the governance process; such forms will not be immediately found, but require experimenting, experiences and joint learning. This implies changing forms of communication and cooperation between science, policy and civil society, which are not yet broadly discussed and developed (Bruckmeier 2013: 266).

Synthesis discourses, knowledge sharing and collective learning become the social practices of knowledge use which help to deal with the complexity of the problems and processes in global environmental governance; this implies more intensive dialogue and communication in the governance process than hitherto.

The interdisciplinary theory discussed in this chapter has, with all its complexity as a knowledge system, the purpose to serve scientific as well as practical purposes. Such a theory is not the last resort and the only intellectual means to cope with the global environmental crisis in environmental governance processes; it is not a theory that needs to be learned by the political actors, as if they should become scientists. The practical utility of the theory is realised through the single arguments it can provide in the governance discourse, arguments about the requirements, the possibilities, and the limits of global environmental governance seen as a part of the broader strategies of social–ecological transformation. This theory is synthesising knowledge that can also be used by transformation action groups, the seminal forms of new collective subjects in the sustainability process. Theories remain controversial in the scientific community and in the political discourse, but the arguments they provide cannot just be ignored. A theory becomes already effective when its knowledge is used by a part of the scientists and other actors involved in the environmental governance discourse. With the social ecological theory discussed here such a bridging discourse between science and governance is already established with the intensifying discussion about social–ecological transformation connected to social and political ecology (Görg et al. 2017; Robbins 2012).

References

Adams, W. M. (1990). *Green Development, Environment and Sustainability in the Third World*. London and New York: Routledge.

Altvater, E. (1991). Universalismus, Unipolarität, Polarisierung: Widersprüchliche Strukturprinzipien einer neuen Weltordnung. *Prokla, 84,* 345–367.

Brand, U., & Wissen, M. (2017). *Imperiale Lebensweise. Zur Ausbeutung von Mensch und Natur im globalen Kapitalismus*. München: oekom verlag.

Bruckmeier, K. (1994). *Strategien Globaler Umweltpolitik*. Münster: Westfälisches Dampfboot.

Bruckmeier, K. (2013). *Natural Resource Use and Global Change: New Interdisciplinary Perspectives in Social Ecology*. Houndmills, UK: Palgrave Macmillan.

Bruckmeier, K. (2016) *Social–Ecological Transformation: Reconnecting Society and Nature*. Houndmills, UK: Palgrave Macmillan.

Electris, C., Raskin, P., Rosen, R., & Stutz, J. (2009). *The Century Ahead: Four Global Scenarios*. Boston: Tellus Institute.

Gerst, M. D., Raskin, P. D., & Rockström, J. (2014). Contours of a Resilient Global Future. *Sustainability, 6*(1), 123–135.

Görg, C., Brand, U., Haberl, H., Hummel, D., Jahn, T., & Liehr, S. (2017). Challenges for Social–Ecological Transformations: Contributions from Social and Political Ecology. *Sustainability, 9*, 1045. https://doi.org/10.3390/su9071045.

Gross, M., & Heinrichs, H. (Eds.). (2010). *Environmental Sociology: European Perspectives and Interdisciplinary Challenges*. Dordrecht: Springer.

Hardin, G. (1968). The Tragedy of the Commons. *Science, 163*(3859), 1234–1248.

Hassan, R. (2009). *Empires of Speed: Time and the Acceleration of Politics and Society*. Leiden and Boston: Brill.

Hassard, J. (Ed.). (1990). *The Sociology of Time*. Basingstoke: Palgrave Macmillan.

Horkheimer, M. (1937). Traditionelle und Kritische Theorie. *Zeitschrift für Sozialforschung, 6*, 245–293.

Moore, J. W. (2015). *Capitalism in the Web of Life*. London: Verso.

Moran, E. (2006). *People and Nature: An Introduction to Human Ecological Relations*. Malden and Oxford, VIC: Blackwell.

Norgaard, R. (1988). Sustainable Development: A Coevolutionary View. *Futures, 20*, 606–620.

Robbins, P. (2012). *Political Ecology: A Critical Introduction*. Chichester, UK: Wiley.

Rohe, W. (2015). Vom Nutzen der Wissenschaft für die Gesellschaft: Eine Kritik zum Anspruch der transformativen Wissenschaft. *GAIA, 24*(3), 156–159.

Rosa, H., & Scheuerman, W. E. (Eds.). (2009). *High-Speed Society: Social Acceleration, Power and Modernity*. Philadelphia: Penn State University Press.

Von Wissel, C. (2015). Die Eigenlogik der Wissenschaft neu verhandeln: Implikationen einer transformativen Wissenschaft. *GAIA, 24*(3), 152–155.

Part III

Towards the Future Society: World Ecology and Changing Societal Relations with Nature

8

Global Environmental Change and the Transformation of the Earth System

During the last years, transformation became a new keyword in the discourses of global change, global governance, and sustainable development. In the discussion of transformative education, transformative literacy, transformative action and transformative science the concept is gradually clarified. It is specified in several variants, where the basic meaning is transformation to sustainability which implies transformation of social and ecological systems. The theoretically most elaborate concept for the transformation towards sustainability is "social–ecological transformation", derived from the overarching theory of nature–society interaction discussed in Chapter 7. Social–ecological transformation is a concept bridging the theoretical discourse in social ecology (where the forms of interaction between social and ecological systems are specified for modern society) and the political discourse about global environmental governance (where the strategies for a transformation to sustainability need to be formulated with the help of scientific knowledge). The term helps to transfer new knowledge to the governance practices, knowledge which is filtered out in the selective practices of policy research and ecological research, in empirical research about the implementation of environmental policies and the functioning of

© The Author(s) 2019
K. Bruckmeier, *Global Environmental Governance*,
https://doi.org/10.1007/978-3-319-98110-9_8

ecosystems. More specifically the knowledge surplus created and made applicable with the theoretical synthesis, from which the concept of social–ecological transformation emerges, can be described as knowledge about the modern world ecology and world society.

1. *The world ecology* emerging through the interaction of social and ecological systems is generated through the human modification of nature and ecosystems in the modern capitalist world system; this system generates the maladaptive change through overuse of natural resources and pollution of the environment resulting in global environmental change. Global environmental change, regarding its consequences for societal and ecological systems, is maladaptive change. The systems analyses from which the transformation concept derives identified the systemic and institutional barriers for a transition to sustainability, the perseverance of non-sustainable interaction and coupling between societal and ecological systems, as described in this chapter.
2. *The emerging world society* is generated through the present transformations of modern society driven by globalisation: the global network society with new divides, social structures and relations that emerge through the economic globalisation process and the international—not necessarily cosmopolitan—components of modern society: the digital economy and the digital divide, the unequal distribution of natural resources and of wealth, the emergence of new social classes in the international economy, the asymmetrical power relations between states and the economic world system, the emergence of new political institutions at global levels, through environmental policy and regulation (described in the following chapter).

Transformation and transition are often synonymously used. Differences between the two terms were discussed in the social–ecological discourse (Fischer-Kowalski and Rotmans 2009), however, this referred only to two of several variants: *socio-metabolic or social–ecological transformations* that imply broader changes in social and ecological systems and in nature–society interaction; and *managed transitions to sustainability* referring to social, economic and cultural changes that may have environmental impacts. The further clarification of the terminology happens in the discussion of social–ecological transformation. The following review of the discussion of social–ecological transformation shows the theoretically specified components of the term. The transition can be understood as equivalent to transformation (Fischer-Kowalski 2011), or as a more open and simple term where not all requirements of the transformation are specified through theoretical analysis.

8.1   The Development of the Concept of Social–Ecological Transformation

Social–ecological transformation as an interdisciplinary term helps to identify forms of regulating and modifying the relations between society and nature that appear in global social and environmental change. The aim of transformation, to reduce the negative social and environmental consequences of the dysfunctional interaction of systems, is only possible through the building of a new society and societal metabolism which implies changes of the economic system and mode of production and changes of the non-sustainable forms of life and consumption in modern industrial society. Transformation of systems, institutions and social practices is a way to solve the problems resulting from intended or non-intended environmental destruction, the unsustainable forms of natural resource use that have been identified in ecology, in the social sciences, and in interdisciplinary environmental research.

1. In systems ecology, the transformational changes discussed refer to ecosystems (Olsson et al. 2004), focusing on ecosystem management and social–ecological resilience; with regard to social systems and processes the few and vague ideas discussed include the formulating of goals and visions, social networking, and connecting nature and culture. Later on the authors completed the discussion of transformation to sustainability with the ideas of adaptive governance, complex systems, and general ideas of innovation, knowledge integration, power relations, still giving a reductionist, ecology and

resilience-related view of knowledge about social systems and processes relevant for transition management and governance (Boyd and Folke 2011; Olsson et al. 2014). No further knowledge about the transformation of the societal and economic systems of modern society is produced in this research.

2. In the social sciences the term of the "great transformation" (Polanyi 1944) is an important source of the present transformation debate. The formulation is also used in the ecological discourse (Haberl et al. 2011), when social–ecological transformation is understood as "another great transformation" for which the one analysed by Polanyi, the transformation to the capitalist market economy in England, is a theoretical model. Transformation in this social-scientific variant means transition to a new form of society or mode of production. For earlier social–ecological transformations in human societies the metaphorical term of "Promethean revolution" was introduced by Georgescu-Roegen in ecological economics, to describe societal transformation processes which are more systematically analysed with the connected concepts of socio-metabolic regimes and social–ecological transformation.

3. The term of social–ecological transformation, used in interdisciplinary research, in social ecology (Hummel et al. 2017) and in the political discourse of sustainable development, is described by Brand and Wissen (2013, 2017) as an "umbrella term"; it comprises various shifts and changes in social systems, including political, economic, cultural shifts in reaction to the interacting economic and environmental crises. In the analyses of these authors the concept is connected with that of an "imperial mode of life". The authors describe two variants of transformation as (a) transformation within the limits of the economic world system, aiming at reconciling economic growth and market processes with social and environmental objectives—the conventional debate of sustainable development, in the lines of reasoning of ecological modernisation; (b) transformation as a critical concept discussed in the interdisciplinary ecological discourse, where the underlying patterns of production and consumption, the system properties of the modern economic world system and the forms of natural resource use are analysed. This debate

proceeds in several interdisciplinary fields as social ecology, political ecology, and ecological economics. The authors suggest a connection of political ecology and critical political economy to enhance the debate and the knowledge transfer with the aim to develop "emancipatory strategies" for combatting the economic and environmental crises (see Sect. 8.2).

The theoretical contexts of the concept of socio-economic and social–ecological transformation can be traced back to the seminal work of Polanyi (Box 8.1).

Box 8.1 Polanyi—"The great transformation"

Planyi's book (1944) is description of the transition from pre-industrial to industrial society and economy, with the historical example of the industrial revolution and the establishment of a modern market system in England.

"Great transformation" in this analysis from critical institutional economics is a theoretical concept to analyse the predominance of markets in modern society, its negative social consequences, and the weaknesses of the modern capitalist market economy based on the idea of self-regulating markets. The critique connects as well to the critical political economy of Marx, as to a "substantivist" view of the economy as societal system for meeting material needs of humans—an ecological view. With the critique of the social dysfunctionality of markets in modern society is connected that of rising inequality, the economic divide between the ones that can pay, for which the economy functions and provides services, and an increasing number of poor outside the economy and the market relations that have no access to the goods and services of the market economy. They need to survive in other forms that are no longer discussed in economics; they are directly dependent from ecosystems and their services, and from local subsistence production.

The development of a dis-imbedded economy: connected with Polanyi's analysis of the great transformation is that of the modern market system through a social transformation called "disembedding"; this includes social and economic changes as changing of mentalities and worldviews that spread in modern society, as described in the classical sociological analyses of capitalism by Weber, where the cultural changes are seen in the development of a new rationality that spreads in the economic system and with bureaucratic organisations as forms of authority—instrumental rationality. Beyond this Polanyi saw also environmental damages as a consequence of natural resource use in modern market economies.

Critique of markets in modern society: the market system reduces the economic actors, producers and consumers, to wage labourers and the natural resources to commodities, a critique in line with that of critical political economy where it was developed theoretically with the concept of exploitation by Marx. The commodification and commercialisation of "everything", especially natural resources, land and labour, implies in the long run a self-destruction of society and overuse of natural resources. Relevant for Polanyi's critique of the modern market system are other forms of economy that existed in pre-industrial societies and are described in terms of reciprocity and redistribution—forms of economic relations and regulatory functions that should be re-established in other forms in modern society to prevent its self-destruction. Polanyi described the counter-movements to the economic maladaptation in different forms of social protection (a historical example is the "New Deal" in the United States after the great depression in 1929, other examples are governmental policies and the struggles of social movements, including the labour movement for social protection). The markets connected with long-distance trade of goods were in pre-industrial societies not part of the everyday life and economy; now they become the nexus of the modern society, as trans-national connections of the capitalist economy which direct the development of the whole society, local economy and trade becoming less relevant. The new market economy is not an autonomous system that developed alone and on purely economic mechanisms; it was established through political processes initiating, supporting and accompanying the transformation—the modern state and the modern economy are interconnected societal systems.

Polanyi's critique of the modern economy argues by way of historical analyses and comparison of economic systems in different societies and civilisations, in the heterodox institutional economics which developed in different forms, struggling with the clarification of the broad term of institution in economic processes. Polanyi's version of institutional economics is closely connected with critical political economy, influenced by Marx; the newer critical theories of world system analysis and capitalist world ecology can be seen as continuing with this critical discourse. The differences between the Polanyian concept of great transformation and the newer versions of social–ecological transformation can be seen in a broadening of meaning that implied the analysis of ecological systems and processes in connection with that if the economic system; the ecological aspects were not ignored in Polanyi´s analyses, but not given the systematic and theoretical analysis as in the present debates.

Sources Polanyi (1944), Polanyi et al. (1957)

Although the term transformation is used in scientific analyses and as a scientific concept, it shows in the formulation of "great transformation" the origins from a diffuse metaphorical notion that needs to be elaborated as a new scientific term. When the term adopts another meaning as the original one by Polanyi, related to the development of modern industry and market-based economy, it appears again in a metaphorical form, as "another great transformation", used to develop the idea of a social–ecological transformation towards a post-industrial and post-capitalist society. This long-term process is difficult to conceptualise and to specify; the transformation comprises contradicting processes and influences, happens in the future, is influenced by many unforeseeable and forms of change and innovation; the final forms of a new society remain unknown. The transformation that began with the counter-current processes of economic globalisation and sustainable development requires the creation of new knowledge to build a new society in the process of transformation itself, starting from systems analyses of interacting social and ecological systems in modern society. Progress in understanding of the possible pathways of transformation develops in the process itself, through continuing knowledge creation and communication: through scenarios and envisioning, research, policies as experiments, collective learning and experience, through the involvement of many different actors, in the processes of global environmental governance at different spatial scales, in connected local, regional national and global governance processes. So far were the elements of social–ecological transformation discussed in the prior chapters.

To advance with the theoretical interpretation and clarification of the term social–ecological transformation and its inter-systemic complexity three conclusions can be drawn from the discussion in the prior chapters:

1. *A general conclusion regarding the theoretical analysis and explanation of the global transformation to sustainability:* all social, political, economic and ecological processes of change at the global scale are part of and influenced by the emergent properties of the capitalist world system. Without analysing the systemic mechanism of the modern world system the processes of environmental policy and governance,

economic globalisation, and sustainable development as an ecological transformation process, cannot be sufficiently understood, influenced, regulated and controlled.

2. *The necessity of a further clarification the transformation concept in theoretical and empirical terms*: social–ecological transformation requires specifying the systems that are included in the process, their interrelations, and the forms of transformative collective action. Only with such interdisciplinary analyses becomes the abstract term of transformation historically and socially concrete, can the concept be used in global environmental governance. Earth system science (Biermann 2014), the research about "New Earth Politics" (Nicholson and Jinnah 2016), and about the capitalist world ecology (Moore 2015) add new knowledge facets to the solution of global environmental problems through environmental governance; however, much of the knowledge required is discussed in other interdisciplinary discourses, in human, social and political ecology.

3. *The necessity to contextualise the concept of social–ecological transformation historically and geographically*: To specify the theoretical concept of socio-ecological transformation further, a comparative analysis of earlier transformations of this kind in the history of human societies is useful. In this analysis the further concept of socio-metabolic regimes, developed in social ecology is useful, helping to specify temporal and spatial factors that influence the social–ecological transformation towards sustainability. The earlier history of modes of production and socio-metabolic regimes, and the transitions between these, are mainly analysed in social ecology (summarised in Chapters 1 and 2). The course of economic resource use and modification of nature was since the early human history until today one of continuing intensification, with changes of modes of production that created possibilities for still more intensive use of and overuse of natural resources and modification of nature and of ecosystems. In the social–ecological transformations from the hunter and gatherer societies to agricultural societies and to industrial society, the intensification of resource use reached unprecedented magnitudes. The average *per capita* consumption of natural resources is ten or more times higher in late industrial than in early hunter and gatherer societies.

At some point in time the interaction between societal and ecological systems exceeds the development capacity of both types of systems through overuse of the global resources. Although carrying capacity is an elastic concept that changes through new knowledge, technological innovations, modification of agro-ecosystems and new forms of agricultural and other economic production, the ecological productivity cannot be infinitely increased. The time where final limits of resource use in the earth system are achieved is approaching; this implies that the coming social–ecological transformation is not of the same type as the great transformations earlier in human history, the transition to agricultural and to industrial society. The coming transition includes a conversion of the long trend towards intensification of resource use in human history: de-growth, de-intensification, and reduction of the levels of resource use achieved. Sustainable development is a process of saving and redistribution of social, economic, and natural resources.

Two kinds of limits of resource use determine the gradual changes within a given socio-metabolic regime or mode of production, and in the ruptures and transformation processes from one socio-metabolic system to a new one: social and ecological or natural limits of natural resource use. The identification of social limits requires the analysis of the combined effects of a socio-metabolic regime: regulation of population growth, social division of labour, property rights, economic and political systems (including the infrastructure systems for natural resource use), power relations and their asymmetries, and the sociocultural values and norms that guide the social forms of life and consumption. Understanding natural limits of resource use requires the analysis of social and economic development with regard to the consequences of human use of natural resources for the environment and ecosystems: the effects of environmental pollution, overuse of natural resources, disturbance of functions of ecosystems that happen today at the level of the earth system and in dimensions of exceeding planetary boundaries of resource use. Scarcity and limits of social and natural resources—in terms of population growth, intensified resource use, social inequality, misery and poverty, social and environmental disasters—reached, at a certain point of time in human history and in the evolutionary development of a socio-metabolic system, levels of instability and crisis that triggered the transition to another socio-metabolic regime. This implies further changes than the political and economic revolutions that marked the transition to modern capitalism in the Western countries and the early attempts to build a post-capitalist society in the failed socialist revolutions in the twentieth century. The dynamics of social–ecological transformations in history comprise transitions to new socio-metabolic regimes and "Promethean revolutions".

The transitions in history did not happen everywhere and at global levels, although they cumulated to global effects in the long run. The transition from hunter and gatherer societies to agricultural societies in the Neolithic revolution happened in limited parts of the world, in different, socially isolated areas. After the transition to agricultural societies, hunter and gatherer societies still existed for a long time, until modernity, in many parts of the world. With the spreading of agriculture, enhanced through the development of the first large-scale societal systems in terms of world systems in ancient history, the new socio-metabolic regimes of agricultural societies spread globally, creating the conditions for a further social–ecological transformation, industrialisation. The socio-metabolic regime of hunting and gathering had no global social and political organisation, developing in more or less isolated local communities, found everywhere where humans lived. It's spreading over the globe is bound to the migration of early humans out of Africa and spreading to other continents. This early socio-metabolic regime is more closely connected to the biological conditions of living and surviving, and the forms of natural resource use were influenced more by these conditions than through further social factors that developed gradually in the historical process as culture-specific transformations of nature.

The socio-metabolic regimes of agricultural and industrial societies became global through social organisation and mechanisms like the state, empires, conquering and colonisation of other societies and territories, with one difference. The introduction and spreading of agriculture was successful in all parts of the world, varying with the natural and climatic conditions and forms of social organisation in a specific area. In contrast to that, industrialisation was up to now only partially successful, limited to certain countries and happening in several phases; with the last phase of delayed industrialisation since the end of the twentieth century in the newly industrialising BRICS-countries, industrialisation is still not reaching all counties and national economies, but already reaching or exceeding the planetary boundaries of natural resource use and the consequence, that future global industrialisation becomes doubtful. What became global was the system of modern capitalism, not its specific form of industrialisation.

More than earlier forms of societies and socio-metabolic regimes, the industrial regime is historically seen as a short-term socio-metabolic regime. It is probably the first socio-metabolic regime in human history that cannot spread over the whole globe; neither economically nor ecologically seen seems its globalisation possible because of social and environmental limits to growth. The industrial society is more an extended form of transition to, or search of, another socio-metabolic regime that is in better balance with the natural conditions of resource use, production and consumption. With the culmination of population growth, resource scarcity, social inequality, and maladaptive change of the industrial system emerge the processes of global social and economic change where the global economy approaches since the end of the twentieth century the physical and biological limits to growth. According to the calculation of the global ecological footprint, one of several indicators for resource use and overuse, the global levels of resource use by humans have exceeded since about 1980 the global bio-capacity; since that time every year more natural resources than can regrow in that time are used. This is not an exact and exclusive indicator for the limits of natural resource use (further indicators are discussed in Bruckmeier 2013: 156ff). However, it gives a first idea how limits of global resource use can be calculated.

The social and ecological limits of natural resource use become core themes in the future debate on social–ecological transformation of the modern world system, the transition to a post-industrial and post-capitalist society. The different indicators of natural limits to growth and resource use are not easily combined with each other and not with indicators that show the institutional and societal limits of the modern economic world system. To develop a system of indicators of limits to natural resource use requires further theoretical analysis and empirical research; so far most indicators are controversially discussed in the scientific and political discourses about the environment. Difficult is the discussion of natural limits—whether there are absolute limits to growth and resource use, or whether the limits change continually with the development of modern society, with new research and technologies that allow, for example, increasing yields in modern agriculture, now through genetic modification of plants and animals. In large parts of the debate about a post-industrial society the environmental problems and the limits to growth are not or insufficiently discussed. In sociology where the term of post-industrial society was created by Bell, it was not seen as a transformative but as an evolutionary process, referred only to continuous structural change in the modern economic systems in advanced and developed countries: sectoral changes in modern economies with a reduction of the industrial and an increase of the "third sector" of service economy. The term of a post-industrial society is now, with the socio-ecological discourse, transformed into another theoretical concept, a transformation of the capitalist society and economy to be achieved through social–ecological transformation.

8.2   Present Discussion of Social–Ecological Transformation

The interdisciplinary discourse of social–ecological transformation is a rapidly developing through the debates in political economy, social and political ecology (Foster et al. 2010; Brand and Wissen 2013, 2017; Brie 2014; Brand 2015; Hummel et al. 2017). Görg et al. (2017) summarise, based on further analyses, the debates about social–ecological transformation as a complex process to develop a more systematic, interdisciplinary and theoretically based analysis. They begin with the diagnosis of different and contradicting transformation processes that happen today simultaneously in society and economy. In social–ecological transformations towards sustainability transformative knowledge is required to intervene in social, political, economic, institutional and technological transformation processes that continue simultaneously.

A better understanding of the societal dynamics, of power relations and conflicting and antagonistic societal processes of change that hinder sustainability is required at first. Then, a better integration of analytical perspectives is necessary to understand transformations in the societal relations with nature, and finally a critical reflection of normative considerations of a desirable goal of global transformations towards sustainability (this especially to improve the political-strategical aims of transformation research). The development of such an integrated, theory-dependent perspective is confronted with a series of difficulties (Görg et al. 2017: 4ff):

- *Ambiguity and disagreement about the meaning and function of the concept*: sustainability transformation is often simplified to a reduction of carbon dioxide emissions, or a low carbon society and economy; furthermore, the transformation is often reduced to political-strategic approaches.
- *Controversies about the definition and accounting for biophysical constraints and limits to natural resource use*: whereas scientifically defined limits are important to determine thresholds of global environmental change, boundaries cannot be defined independent from societal and political processes through which they are influenced and shifted; they are also linked to norms and values that direct processes of resource use and management.
- *Controversies about the subjects or drivers, objects, scope and pace of transformation*: controversial is, whether these are technical or social innovations, or whether the state has a decisive influence, or whether the transformation process requires broader perspectives including interest structures political and economic, power relations, hegemonic constellations in the international system, and their changes. Finally, the clarification of the controversies requires analyses of the societal systems, the societal relations to nature, and the interactions between societal and biophysical processes.

- *In the multi-scale processes of transformation appear tensions and conflicts*: between long-term global transformation processes and further transformations at lower spatial, temporal and social scales; the dominant spatial scale of transformation seems to be regional and national, to which the action of states and governments is oriented. The transformation debate is focused on climate change from which time pressure and urgency of action derive, whereas in the broader discourse of the anthropocene further temporal scales, beyond the institutional processes and capacities appear, finally requiring complicated balances of short- and long-term processes and impacts.

From their analysis and discussion the authors draw the following conclusions about the social–ecological transformation processes: (a) Current societies are in themselves unstable and crisis-driven; it is an open question whether existing resource use patterns can be re-regulated in a sustainable way. (b) Resource use patterns evolve over long periods, require themselves power, and are in complicated ways dependent form existing power relations and hegemonic constellations; a relational perspective is required for a critical concept of transformation. (c) Social-ecological transformations are in conflict with and meet resistance from other, unsustainable transformations; further processes and the systemic relations involved require conceptual and empirical work to allow for improved strategies of transformation to sustainability. (d) A critical approach of social–ecological transformation requires a consideration of the interplay of change and persistence, critical developments, ruptures and discontinuities, the interplay of several transformation processes, and the contexts and wider effects of these.

The critical discussion of social–ecological transformation in broader societal and ecological perspectives is not yet in an advanced state. Many of the reflections by Görg et al. (2017) summarised above leave the impression of being preliminary, more a programme of knowledge integration than its elaboration; the discussion made visible the difficulties of the further discourse of global environmental governance and sustainability transformation. In the transformation discourse much more knowledge needs to be reviewed and used than hitherto: knowledge about systemic processes based on critical systems analyses of social and ecological systems, and knowledge about lifeworld-bound processes of identity building, forms of life and consumption (Ingalls and Stedman 2017; Brand and Wissen 2017). The process of transition to sustainability is much more complex than discussed in large parts of the scientific and political discourse about sustainability, even in sustainability science; the transition cannot be reduced to a policy process. After several decades of discussion the real difficulties of sustainability transformations begin to be understood as including transformations of politics and economy, of society at large, of society–nature relations, and of nature and ecosystems. The combination of social and ecological knowledge becomes decisive for the success of environmental governance, but how to combine different perspectives, theories and approaches is still to clarify epistemologically and methodologically. The sustainability discourse cannot continue in the older forms, with simple ideas and visions, without empirical and theoretical substantiation, and without new knowledge practices based on knowledge integration and synthesis. The newly emerging transformative science, a second variant of sustainability science, can help to pave the way for new forms of knowledge bridging, integration and sharing, in continued controversy with the conventional disciplinary views of scientific research. Much available knowledge for the development of global environmental governance and sustainability transformation, scientific and other knowledge, is not yet used; in the further sustainability discourse it seems impossible to ignore this knowledge.

8.3   The Future Discussion of Social–Ecological Transformation—Conclusions

To realise a difficult project as the renewal and improvement of global environmental governance through strategies of social–ecological transformation, it will at first be necessary to develop the strategies for knowledge integration and synthesis. This requires attempts to connect several discourses that are so far disconnected:

- the critical discussion about the development of "transformative science" (von Wissel 2015),
- the ecological research and discourse about Earth system governance (Biermann et al., see Chapter 4)
- the social–ecological research and discourse of social–ecological transformation discussed here,
- the research and critical debates in political economy, world systems theory, and capitalist world ecology (Foster et al. 2010; Moore 2015; Greffrath 2017), and
- the research and critical discussion of ways of life and consumption and their change, of the creation of social and ecological identities of citizen and consumers (Dauvergne 2008; Ingalls and Stedman 2017; Brand and Wissen 2017).

Within the discourse of social–ecological transformation several variants and relevant forms of research need to be connected (see Table 8.1).

Not all of the discourses and knowledge fields for environmental governance can easily be connected, although they follow similar cognitive interests and goals; the differences and controversies become visible in the processes of knowledge integration, when contradicting knowledge needs to be dealt with, when theoretical, empirical and normative knowledge, scientific and practical or local knowledge need to be connected. In the summary of Moore (2015) the themes and knowledge fields to connect are described as finance, climate, food, work, asking how the crises of the twenty-first century are connected. Moore argues that the sources of today's global turbulence have a common cause: capitalism as a way of organising nature, including human nature. From environmentalist, feminist, and Marxist thinking, Moore develops a theoretical synthesis of capitalism as a "world-ecology" that organises economic wealth, political power and the exploitation of nature. Capitalism's strength and simultaneously the reason for its problems is the capacity to create "cheap natures": labour, food, energy and raw materials. With the planetary boundaries of resource use reached, the shifting of problem-solving to the future and future generations is now in question.

Table 8.1 Theoretical   components   and   variants   of   social–ecological transformation

From the research and knowledge for integrated approaches to social–ecological transformation towards sustainability five areas of research are relevant for multi-scale governance:
1. *Local research on natural resource governance*: common pool resource management, nested systems (Ostrom et al.)
2. *Interdisciplinary social–ecological research*: adaptive management, adaptive governance, sustainability science, resilience research
3. *Interdisciplinary research on the capitalist economy in the social sciences*: critical political economy, accumulation regimes, political ecology, ecological economics
4. *Interdisciplinary theoretical research*: social–ecological system analyses, modes of production, societal metabolism, colonisation of nature, societal relations with nature
5. *Policy-related (applied) research*: international regime theory, global environmental governance, Earth system governance
The emerging picture of social–ecological transformation is that of a complex, multi-scale process that requires to take into account for purposes of governance and regulation a variety of incoherent, conflicting and countervailing processes of social, political, economic and ecological changes and transformations in coupled social and ecological systems. In the further discussion, it becomes necessary to distinguish with the help of theoretical analyses between processes that affect social–ecological transformation and processes that are part of social–ecological transformation.

Sources own compilation

The analysis and the arguments for knowledge integration are convincing, corresponding with other critical theoretical approaches discussed here. But the cognitive and epistemic problems with knowledge synthesis and application are not yet solved. With the discussion of knowledge integration and the formulation of knowledge strategies for global environmental governance it is necessary to leave the "protected communities" of scientists and engage in forms of knowledge bridging, integration and communication between, science, policy and governance, and other practical discourses. Moore's analysis and terminology show, that this will not be easy for the theory of capitalist world ecology. His reasoning is bound to a specific form of theoretical arguments and explanations, conveying much of the theoretical reasoning of Marx, but in a language and terminology that is rather foreclosing than opening for further interdisciplinary knowledge integration and discussion with other theories and approaches. When the different discourses mentioned above, and knowledge from further ones, can be connected in new forms of "theoretical triangulation", where each of the theories and approaches connected compensates weaknesses of the other, some progress in collective learning for environmental governance and development of new strategies for social–ecological transformation can be expected. Such connections of theories, when they include knowledge from the social and natural sciences, cannot be developed in the form of a monolithic synthetic theory, subjecting all synthesised knowledge under one dominant terminology and explanatory logic. More open forms of connecting theories, approaches and specialised knowledge are required for social–ecological transformations at different scale levels; these forms of knowledge integration include interdisciplinary and plural theory construction and synthesis, theoretical triangulation, and combinations of heterogeneous theories (Bruckmeier 2016).

Inter- and transdisciplinary knowledge production and syntheses as new social forms of knowledge generation react to the progressing specialisation and fragmentation of scientific knowledge production, insufficient to deal with complex environmental problems. The limitations and the selectivity of disciplinary and specialised research provide general arguments for interdisciplinary research, which requires further methodological development. Global environmental governance as a new field of policy, where not much practical knowledge and experience with the management and regulation of the complex processes developed, suffers from a series of difficulties. The predominance of natural-scientific knowledge makes the knowledge used in the policy process more difficult, requiring various forms of knowledge bridging and integration. The consequences of specialised knowledge production and use in the natural and social sciences were also discussed as forms of a "*deformation professionelle*" of specialised researchers and experts that cause their selective views and forms of knowledge use. For a long time natural scientists in environmental research neglected social-scientific knowledge, research and theory, assuming they have little relevance for the solving of environmental problems; they described the organisation of regulatory and governance processes with short-cut reasoning of the kind that ecological, biological, physical data can be directly used in the policy and decision processes, attributing the natural sciences a status of authoritative knowledge for the environmental governance processes. For social scientists equivalent forms of neglect of natural-scientific knowledge can be found; for example, conceiving environmental problems as only subjectively and differently constructed by social and political actors with different worldviews and interests is a short-cut reasoning with the consequence of selective knowledge use. Both misleading forms of knowledge search for environmental governance need to be discussed in the governance discourse, because much of the knowledge used comes from specialised research and needs to be put together and configured anew for the governance processes. The development of interdisciplinary knowledge cultures in science is not advanced; much of the learning and development of knowledge integration happens in the processes discussed here, in environmental governance, sustainable development, and social–ecological transformation. Strategies for social–ecological transformation need to be developed with regard to the social and ecological contexts where contradicting and clashing changes and transformations happen simultaneously. Social–ecological transformation is part of broader processes of development, change and transformation in the globalising society. This requires dealing with many contradictions, conflicts and controversies, in science and in politics and governance.

The knowledge created in social–ecological research shows that the further discussion about planetary boundaries and natural limits to growth can be enhanced through the research about socio-metabolic regimes and transitions between such regimes, as discussed in prior chapters. This is the most important contribution from the social–ecological discourse. With regard to the reorganisation of global environmental governance, it is evident that long-term strategies and temporal perspectives are necessary, suggested in the sustainability discourse and in the discussion about social–ecological transformation. However, the difficulties of social–ecological transformation cannot be understood and solved only through political and governance processes. They require further changes in society, in social and collective action, and in systemic processes of connected social and ecological systems that cannot be directly influenced through governance or managed. Although the governance process implies a broadening of the political strategies and forms of action and participation, it can only influence part of the complex processes of social and environmental change that continue, simultaneously with sustainability policies. The ways towards a future sustainable peer-to-peer society where the citizen are assumed to become the decision-makers and produce knowledge together in different groups of knowledge producers, including scientists, and knowledge bearers (Wildschut 2017), are still long, becoming indistinct in the future. But the systemic problems discussed here cannot be shifted in the hope for better solutions in the future. Managing the complexity of the connected societal and ecological systems at global scale requires long-term views and perspectives for multi-scale transformations to sustainability; successive solutions of the problems and progress towards sustainability can be achieved only in long transformation processes.

References

Biermann, F. (2014). *Earth System Governance: World Politics in the Anthropocene*. Cambridge, MA: MIT Press.

Boyd, E., & Folke, C. (Eds.). (2011). *Adapting Institutions: Governance, Complexity and Social-Ecological Resilience*. Cambridge, UK: Cambridge University Press.

Brand, U. (2015). Sozial-ökologische Transformation als Horizont praktischer Kritik. In D. Martin, S. Martin, & J. Wissel (Eds.), *Perspektiven und Konstellationen kritischer Theorie*. Münster: Westfälisches Dampfboot.

Brand, U., & Wissen, M. (2013). Crisis and Continuity of Capitalist Society-Nature Relationships: The Imperial Mode of Living and the Limits of Environmental Governance. *Review of International Political Economy, 20*(4), 687–711.

Brand, U., & Wissen, M. (2017). *Imperiale Lebensweise. Zur Ausbeutung von Mensch und Natur im globalen Kapitalismus*. München: oekom verlag.

Brie, M. (Ed.). (2014). *Futuring: Perspektiven der Transformation im Kapitalismus und darüber hinaus*. Münster: Westfälisches Dampfboot.

Bruckmeier, K. (2013). *Natural Resource Use and Global Change: New Interdisciplinary Perspectives in Social Ecology*. Houndmills, UK: Palgrave Macmillan.

Bruckmeier, K. (2016). *Social-Ecological Transformation: Reconnecting Society and Nature*. Houndmills, UK: Palgrave Macmillan.

Dauvergne, P. (2008). *The Shadows of Consumption: Consequences for the Global Environment*. Cambridge, MA: MIT Press.

Fischer-Kowalski, M. (2011). Analyzing Sustainability Transitions as a Shift Between Socio-metabolic Regimes. *Environmental Innovation and Societal Transitions, 1*(1), 152–159.

Fischer-Kowalski, M., & Rotmans, J. (2009). Conceptualising, Observing and Influencing Social-Ecological Transitions. *Ecology and Society, 14*(2). http://www.ecologyandsociety.org/vol4/iss2/art3/.

Foster, J. B., Clark, B., & York, R. (2010). *The Ecological Rift: Capitalism's War on the Earth*. New York: Monthly Review Press.

Görg, C., Brand, U., Haberl, H., Hummel, D., Jahn, T., & Liehr, S. (2017). Challenges for Social-Ecological Transformations: Contributions from Social and Political Ecology. *Sustainability, 9*, 1045. https://doi.org/10.3390/su9071045.

Greffrath, M. (2017). *RE: Das Kapital. Politische Ökonomie im 21. Jahrhundert*. München: Kunstmann.

Haberl, H., Fischer-Kowalski, M., Krausmann, F., Martinez-Alier, J., & Winiwarter, V. (2011). A Socio-metabolic Transition Towards Sustainability? Challenges for Another Great Transformation. *Sustainable Development, 19*, 1–14.

Hummel, D., Jahn, T., Keil, F., Liehr, S., & Stiess, I. (2017). Social Ecology as Critical, Transdisciplinary Science—Conceptualizing, Analyzing and Shaping Societal Relations to Nature. *Sustainability, 9*, 1050. https://doi.org/10.3390/su9071050.

Ingalls, M., & Stedman, R. (2017). Engaging with Human Identity in Social-Ecological Systems: A Dialectical Approach. *Human Ecology Review, 23*(1), 45–63.

Moore, J. W. (2015). *Capitalism in the Web of Life*. London: Verso.

Nicholson, S., & Jinnah, S. (Eds.). (2016). *New Earth Politics: Essays from the Anthropocene*. Cambridge, MA: MIT Press.

Olsson, P., Folke, C., & Hahn, T. (2004). Social-Ecological Transformation for Ecosystem Management: The Development of Adaptive Co-management of a Wetland Landscape in Southern Sweden. *Ecology and Society, 9*(4), 2. http://www.ecologyandsociety.org/vol9/iss4/art2/.

Olsson, P., Galaz, V., Boonstra, W. J., & Wiebren, J. (2014). Sustainability Transformations: A Resilience Perspective. *Ecology and Society, 19*(4), 1. https://doi.org/10.5751/ES-06799-190401.

Polanyi, K. (1944). *The Great Transformation*. New York: Farrar & Rinehart.

Polanyi, K., Ahrensberg, C. M., & Pearson, H. W. (Eds.). (1957). *Trade and Market in the Early Empires: Economics in History and Theory*. New York: The Free Press.

von Wissel, C. (2015). Die Eigenlogik der Wissenschaft neu verhandeln: Implikationen einer transformativen Wissenschaft. *GAIA, 24*(3), 152–155.

Wildschut, D. (2017). The Need for Citizen Science in the Transition to a Sustainable Peer-to-Peer Society. *Futures, 91*, 46–52.

9

Beyond Globalisation: Another Transformation of the Economic World System

Globalisation of the modern economic world system is one of several social transformations that happen simultaneously with the process of social–ecological transformation towards sustainability. Globalisation and sustainble development can be seen as framing processes, aiming to redirect the development of economic production, distribution and consumption that are part of both of them. The relations between the framing processes are contradicting; furthermore, each of them includes incoherent components. Economic globalisation as the process to create new "development space" for the capitalist world system drives the maladaptive development path of economic growth and industrialisation faster towards environmental and economic crises or collapses. In times where the external limits of natural resource use, planetary boundaries, are reached, the growth path seems more and more risky, economically and environmentally. No spatial extension of the system through conquest and colonisation of new areas on land is possible; key resources of the economy are exhausted; it is always more difficult to find or to exploit new fossil energy sources. Technologies as deep sea drilling of oil and fracking, to gain oil from hitherto economically or technically not exploitable sources, seem dead-end strategies. The conquest

© The Author(s) 2019
K. Bruckmeier, *Global Environmental Governance*,
https://doi.org/10.1007/978-3-319-98110-9_9

of the last economically unoccupied space of the Earth system, the deep sea and the arctic zones, creates only temporary postponement of the exhaustion of non-renewable resources.

9.1   The Contested Concept of Globalisation and the Processes of Global Social Change

The two global processes of globalisation and sustainable development are diffuse with regard to their normative goals and guiding ideas, the concepts used for their analysis, and with regard to their social, economic and ecological consequences. The relations between these two processes, the new wave of globalisation that started in the 1970s with the deregulation of markets, and the sustainability process that started in the 1990s, need to be clarified further. Controversies about their relations and effects continue with different arguments from ecology and economics. However, the growth-related logic of economic globalisation (Purdey 2010) contrasts that of socio-ecological transformation towards sustainability. What has been learned about globalisation and its social effects is less from analyses in economics than in critical globalisation research, especially by Sassen (2007) and Robinson (2009), and in the newer theories of society, especially by Beck, Giddens and Castells.

In interdisciplinary globalisation research and globalisation theory (Held and McGrew 2007) the process is described as a multifaceted and complex, neither exact definitions and causal explanations, nor assessments of the intended and non-intended consequences can be given (see Table 9.1). In this interdisciplinary perspective, controversies continue about the nature of globalisation and different forms of globalisation (Kumar 2003). Globalisation appears as a process of changes in complex, adaptive and co-evolving systems of economy, policy, society at large and ecosystems, with nonlinear changes, with unforeseen effects and surprising events. It cannot be strictly limited to an economic process, as the economic changes influence all social systems and processes in modern society. It is difficult to separate the effects and indirect effects of globalisation from other global change processes; as

Table 9.1 Globalisation in different theoretical perspectives

---

1. Definitions and classifications

1.1 *Definitions and descriptions of globalisation as process of global social change*: Hoogvelt (2001), Held and McGrew (2007)—(a) world compression in space and time (Harvey, Giddens); (b) information and network society (Castells) and global communication through the internet; (c) "real time" economy at global level (acceleration); (d) transnational companies as dominant actors; (e) internationalisation of the state (EU, UN; global governance); (f) globalism as social action (global thinking and action, global movement and migration, global citizenship)

1.2 *Formal classifications of globalisation processes*: Held and McGrew (2007) discuss three hypotheses about globalisation—(a) hyperglobalisation thesis, (b) sceptical thesis, (c) transformationalist thesis that try to identify the nature of globalisation, but obviously it cannot be identified at this level of abstraction and with a diffuse concept of globalisation; the analytical framework and concepts they use, are somewhat more concrete, but remain in the logic of formal and abstract conceptualisation: (a) spatio-temporal dimensions—extensity, intensity, velocity, impact; (b) organisational/social dimensions—infrastructure, institutions and power relations, stratification, modes of production/interaction/communication. With the framing concepts are described different types of thick, diffused, expansive or thin globalisation. Whereas the spatio-temporal relations are only described in certain formal aspects, not well connecting with the social dimensions, the latter include more concrete aspects of political and economic processes that affect globalisation. However, in this form of globalisation theory systemic processes of modern capitalism and its interaction with nature in coupled social–ecological systems are not sufficiently accounted for.

2. Forms and views of globalisation

2.1 *Economic globalisation*: James and Gills (2007), James and Patomäki (2007), James and Palen (2007), James and O'Brien (2007)—empirically described as deregulation of markets, as global flows and movements of goods, capital, technology, services, information, and people (labour), global supply chains (the phenomena and processes since the 1970s); other, non-economic forms of globalisation can be seen as connected with economic globalisation and partly overlapping with this.

2.2 *Cultural globalisation*: Robertson (1992), Appadurai (1996), Pieterse (2003), Rantanen (2005), Flew (2007), Movius (2010); "the compression of the world and the intensification of consciousness of the world as a whole" (Robertson 1992: 8); related to cultural processes and products—international mobility, travel, migration; global communication and mass media, global consumption patterns (food and other), ideologies of globalism (James and Steger 2010).

(continued)

Table 9.1 (continued)

2.3 *Political globalisation*: the growth international and global political systems and processes in manifold forms—international agreements and regimes in different policy fields, internationalisation of the state, international and intergovernmental organisations, international governance systems (as global environmental governance); globalisation of politics—global movements; not developing towards a homogeneous political system, bit constituted through different types of institutions; national states, international organisations and regimes, non-governmental actors and institutions—global governance, global citizenship, global civil society.

2.4 *Social globalisation*: refers to the openness and connections between national states and societies: the impact of globalisation on the life and work of people, families, societies (including employment, working conditions, income, social protection); beyond work-related aspects: security, culture, identity, inclusion or exclusion, cohesiveness of families and communities (World Commission on the Social Dimension of Globalization).

2.5 *Ecological globalisation*: French (2002) refers to the impact of the social and economic processes included in globalisation on the earth system (global environmental governance as containment of economic globalisation—expanding global trade of natural resources is disadvantageous for Southern countries, depriving them of resources that flow in the industrialised countries of the North (Rice: ecologically unequal exchange); strengthening of international environmental institutions is required to convert resource flows and to deal with the consequences of economic growth, unequal exchange and unequal resource use for the global environment.

2.6 *Globalisation as technology-driven innovation and modernisation process*: Archibugi and Mitchie (1997)—technological globalisation is not the end of the nation state and national policies, only changes them through global markets; Archibugi and Iammarino (2002)—globalisation of technological innovation as zip between economic globalisation and raising importance of knowledge in economic processes; Kar and Roy (2015)—globalisation as global integration through technologies.

3. *Systems analyses and systemic theories to analyse globalisation: Theories of capitalism*

(a) *Critical political economy and world system theory*:

theoretical analysis of globalisation as system-specific process of modern capitalism in which the accumulation regimes change, valorisation of nature and natural resources, the restructuring of the centre-periphery division in the modern world system (Hoogvelt 2001, 2006)

(continued)

Table 9.1 (continued)

(b) *Critical sociological globalisation research*:
Sassen (2007), globalisation as "mix of processes"—"growing interdependence and the formation of self-evidently global institutions"—"the
global—whether an institution, a process, a discursive practice or an imagi-
nary—simultaneously transcends the exclusive framing of national states yet
partly inhabits national territories". Sassen's approach is critical with regard
to social changes (Robinson 2009): regarding globalisation from a power
perspective, as multi-scale/level phenomenon, in its historical specificities,
analysing neglected components, but not ecological components of globalisation; two interconnecting changes are dominant in globalisation:
• globalisation as transformation of states through their internationalisation—
  national processes and actors are required for and part of globalisation—
  certain policies cannot easily be denationalised, for example, monetary and
  fiscal policies): selective transformation (only parts of state/governmental
  activities and institutions are transformed to become part of multi-scalar
  global institutions and networks, not everything is globalised)
• globalisation as economic process—the geography of globalisation is different
  from that of world system analysis: not only division of labour between states
  (centre-periphery), but strategic and selective changes—spatial structures of
  globalisation beyond the historical division between centre and periphery;
  part of the global processes embedded in national territories (global cities,
  "Silicon Valleys"); management centre of the global economy lies in the North
  Atlantic region. Globalisation works through different key processes—now
  forms of spatial division at national and global levels, changing functions of the
  state (internationalisation, partial embedding of global processes in national
  processes), changing functions and processes at subnational and local levels
  (horizontal cross-border connections between cities).
(c) *Ecological perspectives*: French (2002)
Time-space compression and connection of distant places through
• economic processes: valorisation and commercialisation of land and other
  natural resources (global trade of resources);
• political processes: changes of states, political systems, political networks,
  global governance systems;
• technology dependent communication—"virtual reality", "simultaneity",
  "acceleration" of communication/interaction;
• socio-ecological processes: human use of natural resources is connected with
  economic and political processes, resulting in changes of land, landscapes,
  ecosystems and their services, industry, settlement, recreation, conservation.
The connection of economic and ecological processes in globalisation: McNeill (2000)
• the socio-economic processes of globalisation in the world system (economic
  growth, global spreading of industrialisation, mass consumption) interact
  with and influence
• the social–ecological processes of environmental change in the earth system
• the feedback from the changes in ecosystems through resource use: exceeding
  the natural limits of resource use and the functions and services of ecosystems.

Sources Own compilation; sources mentioned in the text

globalisation is a fast process of global change, also its consequences and effects change in the longer course of the process. The epistemological and methodological procedures to study such complex systems and processes are not advanced: the dominant approach is systems theory where the reduction of complexity is a heuristic rule for understanding the systems analysed, simplifying the description and reducing theoretical explanation through "dense description" of the structures, functions and processes in the systems. Thus, complexity is used as an argument against causal explanation: too many factors, causes, side-effects and consequences interact to be sufficiently explained; controlled experimenting is impossible, therefore remain few methods to study the complexity: case studies, semi-quantitative models, modelling of systems, and scenario analysis as a form of studying potential future directions of development as consequences of intervening in present changes and redirecting them. These are also methods for studying sustainable development and social–ecological transformation processes.

The different descriptions, interpretations and explanations of globalisation summarized in Table 9.1 show the heterogeneity of specialised globalisation research—and misleading attempts of interpretation and explanation. These include attempts to make a "black box" of changes badly understood from the empirical facts and phenomena into a "grey box" by way of interpreting processes theoretically or through interdisciplinary knowledge integration, in attempts to identify the "true nature" of globalisation. Most of the approaches described above follow such logics of discovery to reveal the nature of globalisation through new research, however, based on unclear assumptions and premises. Only a few approaches, critical political economy, world system analysis, and analysis of social–ecological systems begin the analysis from another diagnosis: connecting studies of the globalisation process systematically to the theoretical knowledge about the modern capitalist world system and its long-term development trends. This seems a more fruitful perspective for analysing globalisation. All of the approaches mentioned describe phenomena relevant for, or connected with, globalisation. Whether there are different processes of globalisation, or whether the different conceptualisations show only the same process from different perspectives, highlighting different aspects, is the core question.

Already around the turn of the century, when globalisation research boomed, Kumar (2003: 110) gave up the idea to clarify the concept of globalisation by way of theoretically defining its true meaning, or identifying the real nature of the process of globalisation through research. He describes an inevitable cognitive circle between normativity and objectivity, prescriptive and descriptive forms of characterising globalisation. There is no absolute and neutral knowledge about globalisation; all theories are imbued with social relations and power relations, and are not independent from the processes they conceptualise and analyse. In this conclusion globalisation is a contestable concept; an inescapable circle of descriptive and prescriptive reasoning in globalisation research prevents to understand the process because of the cultural relativity and perspective-dependence of all knowledge. Different conceptions of globalisation result in different responses and reactions to globalisation; the policy and governance processes reacting to globalisation will be influenced by competing approaches.

This epistemological reasoning about cultural limits of knowledge reduces possibilities of identifying the constituents of globalisation, to understand its social complexity, or finding consensus about its nature; it could in similar form also be applied for the analysis of the other complex processes discussed here, sustainability, transformation and governance. Globalisation is reduced to a non-objective, continually unclear, insufficiently understood and misunderstood process. The interpretation of the epistemic and cognitive problems of globalisation research seems to be determined through a blurring of the differences between the process to be investigated and the scientific interpretation, conceptualisation and analysis of the process. Regarding analysis of globalisation, it would be necessary to dissolve the vicious circle between descriptive and normative reasoning.

Other conclusions are possible. That most of the approaches described above, except the critical globalisation theories, do not arrive at a point to reveal the systemic nature of globalisation can be seen as a consequence of "strategic misreading" of globalisation (Wallerstein) through insufficient theoretical analysis and reflection. Globalisation is part of a long historical process; an analysis of recent trends and development alone does not reveal the historical specificity and social complexity of the long-term processes, but generates shortcut, partial and fragmented analyses, limited with regard to causal explanation and generalisation, sometimes creating methodological artefacts.

The critical perspective of globalisation theory in the sense of critical political economy and world system analysis results in a view of globalisation as a systemic process of development of the capitalist economic system in different historical forms and phases. Globalisation is in these analyses not limited to the economic system, and not limited to the new phase of globalisation since the 1970s, with the deregulation of markets and the neoliberal project. The global spreading of the modern economy and the development of a world market happened, in several historical phases, since the early sixteenth century with the building of the modern world system; during that time the modern world system went through several transformations that had economic, political and social effects.

In the analysis of this historical process, the "*longue durée*" of globalisation, elements of theoretical explanation emerge that cannot be gained from the study of present processes. A historically specified explanation of globalisation is achieved through a Coxian analysis of historical structures (Cox 1983; Hoogvelt 2006) as temporarily stabilised relations of hegemony and international power that provide, together with the economic system analysis of the mode of production, explanations for successive phases and political projects of globalisation during modernity. In the early phase of the modern world system, the conquest-period, Portugal and Spain succeeded as hegemonic powers in dividing the newly conquered and colonised world between them and controlling the global economy until the end of the sixteenth century. In the subsequent phase, the Netherlands was the hegemon for a limited time, followed by the long hegemony of Great Britain in a rivalry with France that ended after the Napoleonic wars with the undisputed British supremacy during the nineteenth century. This supremacy was replaced in the twentieth century and after the end of colonialism through the control of the world in the rivalry between the two new superpowers: the United States and the Soviet Union. The hegemonic relations after the collapse of East European socialism are less clear, as the role of the last superpower, the United States, is changing. Many of the changes influencing future global relations happen in the former "Third World", the post-colonial Global South, that differentiates now in regional variants of development and new growth centres (as the BRICS-countries), which influence the further course of globalisation. Whether the globalisation phase, since the end of the twentieth century, indicates a new world order in which the asymmetrical power relations, the centre-periphery division of countries and national economies, imperialism and hegemony are vanishing, seems doubtful. Rather it can be assumed that the historical structures, the parameters and forms of development of the capitalist world system are changing, and this socio-economic innovation process is still badly understood.

Regarding the globalisation process since the 1970s, several questions about the connections between globalisation, global social change in the broader sense, global environmental change, global environmental governance, and sustainable development need to be answered. The relations and the interaction between the different processes and their sub-processes are rather complex and not coherent: how far does globalisation support environmental regulation; in which regards are the processes contradicting, in which complementary; is globalisation causing more environmental problems, or helping to solve certain problems? Finally, which scientific knowledge, from which disciplines, is required to answer these questions? The controversies about the explanation of globalisation can be partially solved with knowledge from interdisciplinary globalisation research and knowledge from further social-scientific research about modern society.

Globalisation or transformation of the world system—this question by Wallerstein (2000) opens another, historically specified perspective on globalisation, where the argument is unfolding: globalisation, historical and present, is a process to maintain and extend the growth mechanism that drives the development of the capitalist world system. Economic growth is not an autonomous, conflict- and power-free process that follows only economic laws and processes as they appear in the forms of investing and accumulation, trade and exchange, accounting, creation of profits, income and wealth. The asymmetric political and economic power relations in the capitalist world system do not vanish; with the economic development, new inequalities and social divides are generated. Economic growth is a process maintained through unequal economic and ecological exchange and asymmetrical power relations in violent and oppressive forms of domination on the basis of the divide between centre and periphery in the world system. This divide changes throughout the history of the system, resulting in temporarily stabilised constellations that can be described in terms of regimes—political regimes, international regimes, accumulation regimes, and in the theoretically most complex form of socio-metabolic regimes as the systemic structuring of natural resource use. The global north dominated the global south through unequal economic and ecological exchange between the two spatial parts of the economic world system since the times of conquest and colonialism. Throughout this historical process continued the forms of colonisation and human modification of nature and ecosystems. The human domination over nature includes political, economic and ecological forms, the expansion of production and growth of resource use in the economic world system, connected with the geographical expansion and valorisation of ever larger parts of land and nature (Wallerstein 1997). Insufficiently reflected in earlier theorising about modern capitalism, although known, are the ecological transformations of ecosystems that appear in the twentieth century in forms of global environmental change.

The disagreement about the nature, the forms and the consequences of globalisation do not appear primarily in the economic discourse, but in the social-scientific research about consequences of economic globalisation for other parts of modern society and with the question, whether other but economic forms of globalisation are unfolding. The debates (summarised in Table 9.1) show, that many phenomena in political, cultural and social systems of modern society can also be described as globalisation processes, but the simple fact that social processes become always more global and globally connected is not sufficient to argue for the emergence of a new global society. To structures and forms of a "world society", sociological analyses of the forms and consequences of economic globalisation seem useful. Three forms of a global society are developing in sociology:

1. The network economy of global capitalism (Sassen: global cities and global flows of information, people, resources between them);
2. The world market and the global classes, a transnational proto-society that develops outside the countries and national societies; and
3. The global network society as a technocratic utopia, reducing and simplifying social relations and social structures to such of technology-based communication networks.

The results of the sociological analyses are contradicting. Sociological theories that work with the concept of a modern world society (as that of Luhmann) do not provide sufficient knowledge and arguments that show a transformation of modern society; the world society is taken as a fact. All updating theories of modernity by Beck, Giddens and Mol operating with new concepts of modernisation (see Chapter 3) showed: the theoretical analysis ends before the question of transformation is taken up; it seems to be a question not to be answered in sociological research. The pluralisation of sociological concepts of modern society—risk society, knowledge society, network society—shows only facets of change and partial change in the present state of modern society, that appear as confusing and complex. Transformations of the societal systems, as, for example, intended with the concepts of a post-industrial and post-capital society, seem to be "premature utopias" (Bühl 1983). The contradictions and antinomies constructed through the competing theories of society, for example, the idea that one lives today in different forms of society simultaneously, are more sociological fiction and artefacts resulting from the perspectives and concepts chosen. Theoretical analyses of the historical dis-simultaneity of development, early and delayed industrialisation that could be identified through more systematic theoretical analyses of modern capitalist society are only found in the critical theories of capitalist society. Throughout the development of the modern economic world system, coexisted, until today, different historical forms of society, modes of production, subsistence, cultures, socio-metabolic regimes. Large parts of the human population are not yet living in industrialised countries, although they may have access to industrial products; this says more about the historical reality of the present late modern society than the diagnoses of heterogeneous forms of social and technical change.

It can be concluded from the sociological discourse about globalisation: there is no convincing idea or theoretical concept that shows a transformation of the modern capitalist society. The idea of social–ecological transformation is a project of the future, a specific form of transformation of the industrial society that is still incomplete as a global society; industrialisation is an unfinished process that is approaching the limits of its natural resource base which makes a transformation necessary. What happened since the 1990s, after the collapse of the "Second World", is a global transition process in which only two contrasting macro-processes are driving development in different directions, globalisation and sustainable development. Global social change is since the 1990s, with the sustainability process as a form of "ecological globalisation", established in global policies. The contrasting framing processes of economic globalisation and sustainable development happen simultaneously; they encompass all further forms of social, economic and environmental change that need to be analysed as part of these macro-processes. The relations between the two processes can be seen as a dynamic where the logic of growth and the logic of transformation are in continuous conflict and influence each other.

9.2   The New Capitalist Economy—The Digital Economy as a Form of the Information Society

What can be learned from the theoretical reflections about the information or knowledge society is, that the new phase of development since the 1990s, after the collapse of East European socialism, is (a) predominantly global (overlaying processes at national levels), (b) technology-driven (internet and computer technology, digitalisation of information, genetic engineering) and (c) knowledge or information-driven (applying as well more scientific knowledge in social, political and economic processes, as other forms of knowledge). However, the new phase of development of modern capitalism cannot be reduced to one of these three components, neither in terms of explanation, nor in terms of the practical activities and concrete changes observed.

To understand these processes better, it is necessary to understand their structuration through the systemic mechanisms of production and reproduction of the capitalist world system and its dynamics of development.

Most of the new theories of society discussed in Chapter 3 refer implicitly, not always in outspoken arguments, to the development of an information- or knowledge-based economy, connected with the technological changes in knowledge generation, information processing and data analysis through computers and the internet, in short, the "digital revolution". Powell and Snellman (2004: 215f.) described in their review the emerging knowledge economy with the following characteristics:

- In advanced industrial societies happens a transition from an economy based on natural resources and physical inputs to more inputs of intellectual assets and knowledge, indicated in the increase of patents granted and the development of new industries as information and computer technology and biotechnology.
- The research on the knowledge economy is focusing on knowledge production, less on knowledge dissemination, application and its impacts. Neglected are especially the necessities and contradictions of organisational changes which vary between hierarchy, flexible production and increasing control. Moreover, standard metrics for measuring productivity gains and occupational changes remain unclear (for example, simple upgrading of polytechnic schools to universities and accompanying changes of academic degrees, without significant occupational, economic or social changes).

Critical as this review is with regard to the internal organisational principles and criteria of the transition to a knowledge economy, it is ignoring main aspects that would allow for diagnosing a transition in the sense of developing a qualitatively new form of economy.

The influences of information technologies and the internet on social structures and processes are, after two decades of rapid development of internet-based communication, still unclear. Already in early analyses of social effects of the internet were main factors of change described (DiMaggio et al. 2001: 307), but do not give reasons for a technology-based transformation of modern society, remaining single and separate changes: inequality or the digital divide, community and social capital, political participation, organisations and other economic institutions, and cultural participation and diversity. For each of these inexactly described social structures and processes, the authors identify similar trends that provide the conclusion: the internet does not create a new form of society, even at the more specific levels of communication and communication media in modern society; the internet is not replacing but complementing other forms and media of communication and social action. If the internet and its technology create social change, it is only together with other factors and processes, through strengthening certain trends and possibilities of development and weakening others. With these few observations, it seems possible to doubt transformations of large social systems, structures and processes through the information technology. The internet can be understood as a socially influenced communication network, activated through the social users who also can change the forms of use. However, meanwhile another change of internet use is achieved, that produces more unforeseen than foreseen changes: the internet is since several years more used through other automatic, "virtual users", knowledge systems and programmes, creating an invisible "control society"; the social influences of the internet become more indirect, invisible and uncontrollable.

Social changes through technological communication processes and networks include forms of action and communication in nearly all spheres and social systems, in social, cultural, political and economic systems, for commercial, political, scientific, cultural and other purposes. These processes imply a hitherto less observable social change that could be described as lifting the barriers between the lifeworld and civil society, the public sphere, and the systemic spheres of economy, politics, science.

The internet connects and interacts in many forms with economic globalisation processes. Whether and how the internet changes the values and norms of communication, consumption, everyday social action and social routines, also such that affect the environment, is not yet clear. This discussion is mainly continuing in the internet itself, less in scientific studies, and the results are not coherent. Rattle (2010) could not answer the question, whether the internet changes consumption behaviour towards environment friendly and sustainable forms of consumption; it seems possible to conclude that the internet supports contradicting trends, as well as more sustainable and environment-friendly consumption by some consumers, as more resource depleting consumption by others, and mass consumption of high-tech products with short life cycles. The question can be answered better, when "intervening variables" are studied, such as collective forms of modes of life and lifestyles (Brand and Wissen 2017). In later studies, the discussion of sustainable consumption seems to change more and more towards the interpretation of sustainability as development of technologically guided "smart cities" and "smart" forms of social action.

The internet has not yet spread throughout the global society and reached the majority of the global population. When this has happened in future, it is still not sure, that social change is only or mainly effective through new social networks, mobile media and digitalisation, which seems more the vision of the internet controlling international corporations and internet specialists than of scientists who analyse the internet more critically. The dominant process until today is the quantitative growth through new forms of internet use, as described by Earl and Kimport (2011); they studied forms of digitally enabled social change and internet activism, including social network activities, civil society action, social, political, environmental activism, mobilising and protesting, lobbying, campaigning and power-based communication. As a consequence of the shifting of movement and network activities to the internet, one can diagnose parallel forms of social action, conventional and internet-based, but not a transformation of social movements and civil society action to internet communication.

9.3 Transformation of the Global Economy— "Green Economy"

Antinomies of economic growth that have been discussed in growth-critical research on the economy and on natural resource use (Purdey 2010) create further on difficulties to conceptualise in theoretical terms the social–ecological transformation of the modern world system. The antinomies reappear as problems in different versions of a "greening of the economy": in the strategy of a green economy (Kenis and Lieven 2015) that indicates the consensus since the "Rio + 20"-conference in 2012, and in the critical debates about degrowth and sustainable development. In the post-Rio debates of sustainable development, the necessity became evident that sustainable development needs to be specified and connected with a vision of the economic transformation, to become more effective, more directed, creating new regulation strategies. The green economy-vision, arguing within the limits of the neoliberal globalisation project, as the mainstream variant competes with several other visions in the critical economic and ecological discourse where the theoretical elaboration of the concept of social–ecological transformation is in focus.

The idea of a "green economy" appeared first in a report for the British Government, "Blueprint for a Green Economy" (Pearce et al. 1989). The report should clarify whether there can be found consensus about the definition and a common understanding of the notion "sustainable development", especially the consequences of sustainable development for the further economic growth and for policies supporting economic development and their relation with environmental policies. In the report, the concept of a green economy was not theoretically clarified, remaining a buzzword, and in the further reports by the authors, "Blueprint 2: Greening the world economy" from 1991, and "Blueprint 3: Measuring Sustainable Development" from 1994, the debate was broadened to global problems of economic development, including the effects of climate change, depletion of the ozone layer, tropical deforestation and loss of resources in the Global South. The analyses and debates took up the phenomena but did not find ways to reconcile economy and ecology in a concept of a green economy that meets the interests of governmental and economic actors. Insofar the debate documents the continuing dilemma of the sustainability debate to deal with the problems and consequences of economic growth within the limits of the growth paradigm.

The global economic and financial crisis after 2007 stimulated new debates about a green economy as a way to solve the crisis. This happened without further theoretical and critical analysis of the systemic mechanisms and consequences of economic growth. From the United Nations Environment Programme (UNEP) came ideas to develop policies for combatting the global economic crisis with projects of the greening of the economy that should stimulate the development of a more "green economy" (AtKisson 2012). As the earlier debates about the green economy, the new ones were driven by government- and economy-close institutions and actors. In the newer debates, the tactical reasoning with a misplaced idea became clearer: sustainable development should serve the revitalisation of economic growth in the global economy, stabilising the global economic system without transforming it. The growth paradigm was rethought in terms of "green growth", using a variety of vague ideas that do not show ways to overcome the exclusion of large parts of the global population from the economic growth mechanism, in spite of focusing on poverty reduction. In the "beyond growth", discussion about a green economy initiated by AtKisson (2012) appears a confusing blending of ideas about green growth, genuine savings, sustainable development, gross national happiness, degrowth and other growth-critical, mainly normative ideas, without advancing towards a critical, theory-based review of the systemic mechanisms of economic growth in the modern world system.

Box 9.1 The UNEP strategy of "Green Economy"

In October 2008 started the "Green Economy Initiative" of the UNEP to generate analyses and policies for investment in "green sectors" and in the ecological modernisation of economic sectors with negative environmental impacts. The ideas were framed in a strategy of a "Global Green New Deal (GGND)" in a report from April 2009, suggesting a policy mix for stimulating economic recovery and growth, simultaneously improving the sustainable development of the global economy. Governments should stimulate the funding of green sectors with three targets of economic recovery; poverty eradication, and reduction of carbon emissions and ecosystem degradation; national and international policies should support the green stimulus funding (UNEMG 2011).

In the following debates, the notion of transformation appeared, not clarified further though scientific analyses, again remaining a buzzword in the public and policy discourses under the guidance of the United Nations. At the occasion of the Climate Change Conference in Copenhagen in 2009, the UN supported the idea of a green economy as economic transformation to address the global economic crisis. The debates continued in the UN-UNEP institutions, finally creating the consensus at the "Rio+20"-Conference on Sustainable Development in 2012, where the green economy strategy for sustainable development was a main theme. This process brought numerous publications trying to define and clarify the idea of a green economy under the guidance of UNEP. To make the concept and economic analyses supporting it more credible, UNEP cooperated with think tanks and commercial organisations, making visible the strategy was one for the economic actors and global players where consensus is found about stimulating economic growth as a main goal.

In December 2011, the United Nations Environment Management Group (UNEMG, a cooperation more than forty specialised agencies, programmes and organs of the United Nations) presented its perspective on a green economy "Working Towards a Balanced and Inclusive Green Economy", adopting the definition provided by UNEP in its 2011 Green Economy Report.

The efforts to achieve consensus about a definition of "green economy", if not in the scientific, at least in the political discourse, were partly successful so far; the green economy initiative of UNEP succeeded in finding support of a series of non-governmental organisations for working towards a strategy for a global green economy. The UNEP defines a green economy as "one that results in improved human well-being and social equity, while significantly reducing environmental risks and ecological scarcities. It is low carbon, resource efficient, and socially inclusive" (UNEP 2011). The definition was taken up in other reports by the UNEMG and the OECD. The Green Economy Coalition of non-governmental organisations defines green economy as "a resilient economy that provides a better quality of life for all within the ecological limits of the planet."

The history of the discourse about a green economy is, as the institutions and actors involved show, a debate in the forms and limits of "sustainable development diplomacy" (Moomaw et al. 2017). The concept, the terminology, the analyses show the vague and diffuse ideas and terms that guide the debate; they characterised the sustainability discourse since the beginning. The vagueness can be seen as a reason for its success, although no consensus was created through research and scientific knowledge that was only used selectively and to a limited degree. The consensus about a green economy is a symbolic consensus among political actors about political goals. The whole debate is focused on the ideas articulated by the UNEP, also when scientific task force groups discuss the concept, as in Selin and Najam (2011). The approach has influenced and modified the global environmental discourse, but tends to exclude other approaches and ideas, as indicated in the rhetoric of "our common future" or "humankind's economy". As earlier attempts, for example, the idea of ecological modernisation (Mol), it can be seen as containment action to prevent or absorb alternative ideas of a transformation of the economic system, as they come up but with the discourse of social–ecological transformation. Adopting the term transformation as a new buzzword in the green economy discourse is part of such containment.

Sources UNEP (2011), UNEMG (2011), Selin and Najam (2011), AtKisson (2012), Moomaw et al. (2017)

The limits of the "green economy" ideas are shown by Kenis and Lievens (2015) in a more systematic scientific analysis: it is an attempt to solve the economic crisis and to dissolve the contradictions and antinomies of the capitalist world system within the boundaries of the system, seeking for solutions to environmental problems through market mechanisms. Such an attempt to reinvent capitalism for maintaining the growth imperative faces now growing critique with the discourse of degrowth and other more critical debates as that of social–ecological transformation. Environmental governance is with the green economy ideas reduced to technocratic reforms that are not primarily aiming to solve environmental problems but to spread the illusion that talking in the common interest of humankind is enough to begin to discuss ways to a common future in terms of sustainability. Interpreting the green economy discourse with the critical notion of idea post-politics, Kenis and Lievens criticise the way how nature–society interaction is constructed in the discourse and show how socio-ecological alternatives are foreclosed. The arguments for repoliticising the environmental discourse, for environmental justice and democratic decision-making (Kenis and Lievens 2015: 156ff.) start also from normative ideas, but the debate opens for a critical assessment of the government- and economy-close epistemic community that instrumentalised the green economy debate to bring it away from the system-critical transformation discourse. As other authors in ecological research on the use of natural resources (Ostrom, McCay), they see a necessity of renewing and strengthening the commons as complementary to private property and market relations, with the argument: no society can survive without communing and a minimal form of sharing. This is not yet a theoretically formulated alternative to the capitalist growth paradigm, but at least the beginning of such a discussion, with elements of economic systems that developed during the long history of human societies and need to be rethought for a potential future society and economy where economic institutions complementary to the market need to be developed: forms of commons, of reciprocity, of redistribution and sharing of resources.

The terminology of the green economy debate consists of abstract, diffuse terms and value-loaded formulations. The green economy ideas, presented as an attempt to renew the discourse of sustainable development, do not go beyond the forms of discussing sustainability earlier in the discourse, which is one of the reasons for its lack of success: sustainability is interpreted in normative terms, reduced to a goal for policies, but not analysed with regard to the dysfunctionality of the economic world system, although it is seen that changes of the economy are necessary and politically the action of nation states is not sufficient to achieve sustainability and solve environmental problems. When it comes to the priorities in the contrasting requirements of sustainable development, the economy and its functioning is the top goal (Halle in: Selin and Najam 2011: 21), social and environmental sustainability is less important, to be taken into account only insofar as they can be realised with the economic system. The green economy discourse is striving to repeat the success of the idea of sustainable development in gaining scientists and politicians and other actors engaged in the policy debate—to introduce and spread a new economic terminology and thinking that reacts to the sustainability debate, but follows unspoken premises of strengthening the functioning of markets and revitalising economic growth.

9.4   Growth and Degrowth

The critical scientific and political discussion about the consequences of exponential economic growth and growth of natural resource use are continuing since the 1970s, when the "Limits to growth" report of the Club of Rome was published. Only in the last decade started a broader growth-critical discourse: a new global degrowth movement developed (Borowy and Schmelzer 2017) and the public communication about "the folly of growth" ("New Scientist", October 2008) intensified.

Concrete ideas and suggestions on how to reduce environmentally destructive economic growth are rare and the discussion of possible forms of a transformation of capitalist growth machine to a sustainable post-industrial economy of the future is still at the beginning in science and policy, in economics and in the environmental discourses about greening of the economy and ecological modernisation. A recent breakthrough with regard to societal transformation happened with the discussion about another great transformation or socio-ecological transformation of the global economy. The global economy is since decades passing through a long downward cycle with the usual form of destructive degrowth in the capitalist system through the devaluation of capital, creation of unemployment and wasting of natural resources. The short-term solutions through economic change are not that of degrowth, but of revitalising growth, stimulating the economic growth mechanism again in the hope for a new upward move: to make once more the systemic mechanism working for a partial and temporary solution of the problems it created. Creating new jobs and stimulating consumption are the main objectives; new research and technology development that should help to restart economic growth may then, as a secondary effect, also help to reduce somewhat the damages done to ecosystems and nature.

When the capitalist growth engine is identified as the main problem of further economic development and socio-ecological transformation, it is necessary to analyse how the conflict between economic growth and the environment, discussed insufficiently so far in the scientific and political ecological discourses, can be solved in the longer run through strategies of transformation to sustainability. The arguments of mathematicians and ecologists, the early critical analysts of exponential growth processes, were obviously not sufficient to argue for degrowth, although no convincing arguments for the long-term possibility of exponential growth in whatever form—economic growth, growth of natural resource use, population growth—could be found to reject the theoretical critique of exponential growth. Simply ignoring this critique is not possible in future; it can no longer be exposed as creating phantasies and utopian ideas—the nexus between capitalist growth and disruption of the environment is evident. The situation since the days of the early limits-to-growth debate can be described as the dilemma of knowing about the growth problems, but not knowing how to solve them.

As a consequence, the growth debate has become another example of the trapped discussion of "wicked problems" and "clumsy solutions" (Brooks and Grint 2010). The arguments for possibilities of continued economic growth and its "greening" articulated in different forms in the ecological discourse, are already discussed and devalued: the idea of a "green growth", the ecological "Kuznets curve", ecological modernisation as a safe future of the capitalist economy, creating a new ecological rationality in economic production to phase out the dominant economic rationality, dematerialisation of production and consumption, the "factor 4" or "factor 10" efficiency revolutions through saving of material and energy—all have been devalued through the practical experience that continuing economic growth eats up the environmental improvements and efficiency gains.

The phenomena known since long in economics as the Jevons paradox, the Kazzoom-Brookes postulate, or the rebound effect, indicate the impossibility of system-internal remedies in forms of economic internalisation of external effects. The alternatives discussed above can be seen as idealist, visionary or utopian ideas that cannot be realised in the capitalist system and the growth paradigm: as attempts of self-curation of the system, to reconcile economic growth and the environment without addressing the question of system transformation. This debate continues parallel to the new transformation and degrowth debates. It is, for example discussed, how the ecological turn or conversion of the global economic system can continue that began with sustainability policies, especially in the search for environment-friendly energy systems through the use of renewable energy sources. The arguments already found decades ago in the ecological modernisation discourse come up again as ideas of transforming the growth mechanism: new environmental technologies that help to reduce environmental pollution, to reduce CO_2 emissions in the atmosphere, and to save material and energy. These ideas are not only seen as gradually improving the environment but as the drivers of growth in the global economy for the coming decades as well, for the next long wave (Kondratiev-cycle). In misleading terms, the changes in information technologies are also called a new or "the third industrial revolution", described by Rifkin with the simple message: renewable energies and the internet make the future global economy.

The progress since the early ecological modernisation debate in the 1990s is that of a broadening of the ecological modernisation debate from one of the industrialised countries to the global economy, and the insight, that the critical component in the economic growth process is the energy intensity of the modern lifestyles and the "energy hunger" of the economy and the new technologies. The expectation that with the technological rebuilding of the global economy the polluting branches of industrial production will vanish was not realised. So far the pollution has only migrated to the Global South, to the newly industrialising countries, and the first mistakes with the transformation of the industrial energy system have already been made. To use of solar energy in the form of a large-scale technological system, providing all electricity for Europe by covering the Sahara desert with solar panels, the "Desertec"-project, collapsed quickly for reasons of its incoherent planning and as a growth-adapted technical-fix solution in gigantesque forms of a new "Atlantropa"-project. It showed, furthermore, that new energy technologies require long development phases that are not possible without public funding and subsidies. The main problem is the form in which new environmental technologies are to be developed:

• as part of the established capitalist economy, where the energy sector is an example of an oligopolistic sector, with large international corporations, concentration of capital, and large-scale technological systems; the developing of a renewable energy source as an economic alternative to polluting fossil resource-based technologies is here distorted and devalued through the growth and profit-dependent system mechanism; and
• or in small, decentral and local forms as part of an alternative environment-friendly sector as it is, for example, existing in the economic projects of fair trade or organic farming.

The conflict between globalisation or growth-driven and sustainability- driven forms of organising social processes of change and transformation will continue in the future discourse of global environmental governance. What has been learned already from other forms of using renewable energy resources, especially with the production of bio-energy on agricultural land, is: the development of green energy in the existing structures of the global economy causes social, economic and environmental conflicts, is not a win-win solution. In one way or another, the question of social–ecological transformation of the economy comes back. A similar conclusion can be drawn from the international climate policy: it is not enough to organise carbon trading and decarbonise the economy or develop technocratic mechanisms of climate change mitigation. Necessary is, to enforce discussions about a transformation of the capitalist economy: a system transformation towards degrowth, a planned transition to a non-fossil fuel economy, supported through the development of new forms of democratic decision-making and new normative orders in terms of environmental justice (Storm 2009).

9.5   Conclusion—Growth and Transformation

With the further critical discussion of the relationships between globalisation, economic growth, transition to degrowth strategies and strategies of social–ecological transformation, the discourse of global environmental policy and governance will become more complex and complicated, requiring the development and adoption of new knowledge use practices as discussed in this book.

To deal with the growth problems, it will be necessary to deconstruct the growth question, to differentiate the growth concept and to develop strategies to achieve degrowth that are clashing with globalisation-driven economic development. This implies that different forms of growth need to be discussed separately and their interconnections be reconstructed through more refined forms of system analyses and research. It is no knowledge-generating strategy to use vague ideas of exponential growth and to argue with similarities or analogies between different phenomena of economic growth, the growth of population, resource use, environmental pollution, environmental catastrophes and further forms of growth. Necessary is to identify the systemic mechanism of economic growth and its connections with other forms of growth; this will not be possible without the analyses of interactions between social and ecological systems discussed in the discourse of social–ecological transformation.

The contrasting logics of economic globalisation or growth and social–ecological transformation to sustainability cannot be easily dealt with in global environmental governance; they reach far beyond policy and governance debates, therefore requiring more inter- and transdisciplinary forms of knowledge creation than so far practised in environmental research. However, it can be foreseen, that the continuing degradation of ecosystems, progressing pollution and scarcity of natural resources will create more and more conflicts and problems; reactions to the limits to growth, the social and ecological limits of natural resource use, seem to become unavoidable. These reactions need to deal with questions of the transformation of modern society that reach in the distant future about which no data and no knowledge can be produced. It becomes necessary to develop more sophisticated methods of reflexive science (Popa et al. 2015) and forms of thinking about the future, of anticipation (Groves 2017), of future studies (Schatzmann et al. 2013), of scenarios and of better understanding potential pathways to the future.

References

Appadurai, A. (1996). *Modernity at Large: Cultural Dimensions of Globalization*. Minneapolis: University of Minnesota Press.

Archibugi, D., & Michie, J. (1997). Technological Globalisation or National Systems of Innovation? *Futures, 29*(2), 121–137.

Archibugi, D., & Iammarino, S. (2002). The Globalization of Technological Innovation: Definition and Evidence. *Review of International Political Economy, 9*(1), 98–122.

AtKisson, A. (2012). *Life Beyond Growth*. http://lifebeyondgrowth.org.

Borowy, I., & Schmelzer, M. (Eds.). (2017). *History of the Future of Economic Growth: Historical Roots of Current Debates on Sustainable Degrowth*. London and New York: Routledge.

Brand, U., & Wissen, M. (2017). *Imperiale Lebensweise. Zur Ausbeutung von Mensch und Natur im globalen Kapitalismus*. München: oekom verlag.

Brooks, S., & Grint, Keith (Eds.). (2010). *The New Public Leadership Challenge*. Houndmills, Basingstoke, UK: Palgrave Macmillan.

Bühl, W. (1983). Die 'Postindustrielle Gesellschaft': eine verfrühte Utopie? *Kölner Zeitschrift für Soziologie und Sozialpsychologie, 35*(4), 771–780.

Christoff, P., & Eckersley, R. (2013). *Globalization and the Environment*. Lanham, MD: Rowman & Littlefield.

Cox, R. W. (1983). Gramsci, Hegemony, and International Relations: An Essay in Method. *Millennium: Journal of International Studies, 12*(2), 162–175.

DiMaggio, P., Hargittai, E., Neumann, W. R., & Robinson, J. P. (2001). The Social Implications of the Internet. *Annual Review of Sociology, 27*, 307–336.

Earl, J., & Kimport, K. (2011). *Digitally Enabled Social Change: Activism in the Internet Age*. Cambridge, MA and London: MIT Press.

Flew, Terry. (2007). *Understanding Global Media*. Houndmills, Basingstoke: Palgrave Macmillan.

French, H. (2002). Reshaping Global Governance. In *State of the World 2002*. New York: The Worldwatch Institute and W.W. Norton.

Groves, C. (2017). Emptying the Future: On the Environmental Politics of Anticipation. *Futures, 92*, 29–38.

Held, D., & McGrew, A. (Eds.). (2007). *Globalization Theory: Approaches and Controversies*. Cambridge: Polity Press.

Hoogvelt, A. (2001). *Globalization and the Postcolonial World: The New Political Economy of Development* (2nd ed.). Basingstoke, UK: Palgrave.

Hoogvelt, A. (2006). Globalization and Post-modern Imperialism. *Globalizations, 3*(2), 159–174.

James, P., & Gills, B. (2007). *Globalization and Economy, Vol. 1: Global Markets and Capitalism*. London: W.W. Norton and Sage.

James, P., & Patomäki, H. (2007). *Globalization and Economy, Vol. 2: Global Finance and the New Global Economy*. London: Sage.

James, P., & Palen, R. (2007). *Globalization and Economy, Vol. 3: Global Economic Regimes and Institutions*. London: Sage.

James, P., & O'Brien, R. (2007). *Globalization and Economy, Vol. 4: Globalizing Labour*. London: Sage.

James, P., & Steger, M. (2010). *Globalization and Culture, Vol. 4: Ideologies of Globalism*. London: Sage.

Kar, S., & Roy, S. C. (2015). Globalization of Technological Innovation: Challenges and Opportunities. *Science and Culture, 81*(7–8), 182–186.

Kenis, A., & Lievens, M. (2015). *The Limits of the Green Economy: From Reinventing Capitalism to Repoliticising the Present*. London and New York: Routledge.

Kumar, V. S. S. (2003). A Critical Methodology of Globalization: Politics of the 21st Century? *Indiana Journal of Global Legal Studies, 10*(2), 87–111.

McNeill, J. R. (2000). *Something New Under the Sun: An Environmental History of the Twentieth Century*. New York and London: W.W. Norton.

Moomaw, W. R., Bandary, R. R, Kuhl, L., & Verkooijen, P. (2017). Sustainable Development Diplomacy: Diagnostics for the Negotiation and Implementation of Sustainable Development. *Global Policy, 8*(1), 73–81.

Movius, L. (2010). Cultural Globalisation and Challenges to Traditional Communication Theories. *Platform: Journal of Media and Communication, 2*(1), 6–18.

Pearce, D. W., Markandya, A., & Barbier, E. B. (1989). *Blueprint for a Green Economy*. London: Earthscan.

Pieterse, J. N. (2003). *Globalization and Culture*. Lanham: Rowman & Littlefield.

Popa, F., Guillermin, M., & Dedeurwaerdere, T. (2015). A Pragmatist Approach to Transdiciplinarity in Sustainability Research: From Complex Systems Theory to Reflexive Science. *Futures, 65*, 45–56.

Powell, W. W., & Snellman, K. (2004). The Knowledge Economy. *Annual Review of Sociology, 30*, 199–220.

Purdey, S. J. (2010). *Economic Growth, the Environment, and International Relations: The Growth Paradigm*. New York: Routledge.

Rantanen, T. (2005). *The Media and Globalization*. London: Sage.

Rattle, R. (2010). *Computing Our Ways to Paradise? The Role of Internet and Communication Technologies in Sustainable Consumption and Globalization*. Lanham, MD: Altamira Press.

Robertson, R. (1992). *Globalization: Social Theory and Global Culture*. London: Sage.

Robinson, W. I. (2009). Saskia Sassen and the Sociology of Globalization: A Critical Appraisal. *Sociological Analysis, 3*(1), 5–29.

Sassen, S. (2007). *A Sociology of Globalization*. New York: W.W. Norton.

Schatzmann, J., Schäfer, R., & Eichelbaum, F. (2013). Foresight 2.0— Definition, Overview & Evaluation. *European Journal of Futures Research, 1*(15). https://doi.org/10.1007/s40309-013-0015-4.

Selin, H. & Najam, A. (2011, March). *Beyond Rio + 20: Governance for a Green Economy* (Pardee Center Task Force Report) Boston: Boston University, Frederick S. Pardee Center.

Storm, S. (2009). Capitalism and Climate Change: Can the Invisible Hand Adjust the Natural Thermostat? *Development and Change, 40*(6), 1011–1038.

UNEMG (United Nations Environment Management Group). (2011). *Working Towards a Balanced and Inclusive Green Economy*. Report. http://hdl.handle.net/20.500.11822/8065.

UNEP (United Nations Environment Programme). (2011). *Towards a Green Economy: Pathways to Sustainable Development and Poverty Eradication—A Synthesis for Policy Makers*. Report. www.unep.org/greeneconomy.

Wallerstein, I. (1997, April 3–5). *Ecology and Capitalist Costs of Production: No Exit*. Keynote Address at PEWS XXI, The Global Environment and the World System. University of California, Santa Cruz.

Wallerstein, I. (2000). Globalization or the Age of Transition: A Long-Term View of the Trajectory of the World System. *International Sociology, 5*(2), 251–267.

10

Rethinking and Renewing Global Environmental Governance as Part of Social–Ecological Transformation

In this chapter, the limits of governance are identified and possibilities to improve the knowledge basis of global environmental governance are discussed in two steps, based on the analyses in the preceding chapters. First, environmental governance is connected with social–ecological knowledge, showing how governance strategies can be embedded in interdisciplinary knowledge about global social and environmental change. Thereafter, governance is discussed in terms of the temporal structuring of possibilities and pathways of transformation. The structuring of the long-term transformation to sustainability shows the difficulties of transformative action as part of global environmental governance: it requires intensive cooperation, knowledge integration and synthesis. As a consequence, governance strategies need to be differentiated in temporal, spatial and social dimensions, knowledge practices improved through continuous monitoring, collective learning and updating. The specialised knowledge and expertise in policy science and governance research plays only a limited role in this improvement that includes social–scientific knowledge about the broader contexts of societal development.

© The Author(s) 2019
K. Bruckmeier, *Global Environmental Governance*,
https://doi.org/10.1007/978-3-319-98110-9_10

The deficits of international environmental, climate and sustainability policies cannot be sufficiently explained through institutional inertia, ignorance, asymmetrical power relations, difficulties of negotiations and interest matching, or the influence of powerful political and economic actors and institutions that control the policy processes. Such arguments prevail in global environmental assessments, one of the main knowledge sources for global environmental governance: in the Millennium Ecosystem Assessment, in the IPCC-reports on climate change, in periodical reports as "The state of the environment", even in the critical policy documents of independent, non-governmental institutions like the IAASTD (2009), and in numerous national and international policy documents from environmental movements. The filtering of arguments and knowledge in the processes of policy formulation and implementation passes through many layers of discussion and decision-making in research, public and political debates and controversies, political programming and evaluation. In the knowledge transfer process and the selection of knowledge included, the system analyses as key components of strategies for social–ecological transformation are watered down to create consensus and support through the many scientists, political actors and stakeholders participating in global environmental assessments. The vague goals for stakeholder participation in the assessments, to increase their legitimacy and credibility are not necessarily the best for the quality of the assessments; deliberative policy learning as a more ambitious aim is not always achieved (Garard and Kowarsch 2017). The existence of multiple legitimate perspectives, the necessity to engage in multiple epistemologies, and the multiple roles of science in governance processes—all indications of the complexity of governance—make the knowledge processes contingent upon contrasting interests and aims, do not necessarily improve the knowledge for governance and support quality assurance, can also make it more selective and create uncertainty (Kovacic 2017).

Power relations and knowledge problems in environmental governance are discussed with regard to sustainability and the deficits of integration of social-scientific knowledge and citizen or lay knowledge in governance processes (Peterson 2018). Power relations in the international policy arenas depend on many further factors outside the politically organised spheres of national and international policy processes. Improvements in governance do not only depend on organisational structuring or on the broadening of institutions, actors and decision-making that happened with the governance concept. *The generation and utilisation of further knowledge about the processes to be governed, controlled and navigated—complex global social and environmental change—is more important for the success of governance.* Environmental governance depends from processes in social, ecological and social–ecological systems which never before have been subjected to governance and political regulation.

With the complexity of the process and the context of governance arise either questions of how can complex systems and processes be managed, their complexity reduced, taking into account the limited number of factors that can be controlled or the amount of information used in a governance process? (Becker and Ostrom 1995; Agrawal 2003); or questions how can the complexity of governance increase, how can governance systems learn when multiple perspectives need to be taken into account? (Boyd and Folke 2011; Kovacic 2017). The global scope, the many actors involved, and their heterogeneous, competing, or vested and unclear interests and forms of influence limit the effectiveness of multi-scale governance processes. In the final analysis, everyone is a stakeholder in global environmental governance, where not everyone can be involved actively through participation; participation remains selective. The complexity is dealt with global governance and environmental regimes in simple ways through the participation of stakeholders: global environmental governance begins under conditions of the existing, insufficiently developed institutions, mainly government-dependent ones, and proceeds to a limited broadening of the institutional setting through participation and co-optation of nongovernmental organisations and other stakeholders, with unclear forms of participation and legitimation. Adaptive institutions that allow the different actors to deal with complexity, uncertainty and different forms of change (Boyd and Folke 2011: 3) are one way of dealing with complexity, especially that of ecosystems to which adaptive management is oriented, not necessarily that of the social systems connected with them.

Bown et al. (2013) find it more difficult to assess adaptive management and co-management because of the necessity to balance ecological impacts with that of participation, social and community issues.

In environmental governance, confronted with this situation and impermeable complexity, the responses and reactions cannot be to include ever more actors and knowledge in the governance process, but to restructure—on the basis of collective learning and experience—the organisational and the knowledge use processes, critically reviewing the knowledge used, and broaden knowledge use in methodologically controlled ways. The reorganisation of knowledge processes as the first step of renewing global governance is a form of building of epistemic capacity; this is described in this chapter in terms of social and ecological requirements, of the temporal and spatial structuring of the governance processes.

10.1 Knowledge Selection in Policy and Governance Processes

The knowledge selection, the forms of assessment, and the use of multiple knowledge forms and knowledge syntheses for environmental policies and governance processes need to be evaluated in the processes. Although the selective information in policy processes was intensively studied (Fischer 2000; Dimitrov 2006; Lemos and Agrawal 2006; Ascher et al. 2010; Atkinson et al. 2011; Young 2011; Prewitt et al. 2012; Jones et al. 2012; Sörlin 2013; Cornell et al. 2013; Daviter 2015; Millner and Ollivier 2016), scientific or political consensus about policy improvements has not been achieved. Instead of working in the long-term perspective and dealing with that what cannot be planned, managed, and dealt with in terms of positive knowledge, the system complexity and the unknown future, the discussion continues about how to make the policy and governance processes more effective and efficient—as if one would know already how to achieve the unknown new forms of sustainable society and economy. Two forms of limits of governance, agency, and scientific knowledge are to be reflected in analyses of governance processes:

- *final limits* of agency and knowledge (regarding that, the distant future and how to deal with it are critical questions), and
- *social limits* resulting from the social organisation of governance and selectivity of knowledge production and use, the normative framing of policy processes, and the interests and negotiations between governmental and further actors (regarding that, the governance process appears as continuously fluctuating and changing).

The knowledge selection and negotiation processes in the science–policy interaction become visible with inter- and transdisciplinary research; for environmental policy new forms of cooperation in knowledge creation and use need to be sought. Simultaneously, and in contrast to this broadening, the knowledge bases for environmental governance, the knowledge filtering, assessment and selection to support decision-making becomes technocratic and commercialised, a general trend connected with the internet use, including knowledge assessment through software and search programmes, for example, for assessing knowledge quality by distinguishing in binary code-form "high quality" from "low quality" information (Bizer and Cyganiak 2009).

An important component of knowledge filtering in environmental policy processes is visible in the selective communication of underlying mechanisms, reasons and causes of environmental destruction—altogether: the system analysis of coupled social–ecological systems. Even when it is acknowledged, that "the elephant in the room" (Newell 2011) is modern capitalist and industrial society as large-scale system, causing environmental destruction through the mode of production and economic organisation of resource use, sustainability policies have hardly taken up knowledge from societal system analyses since the begin of the discourse three decades ago. This began only in recent years (Martinez-Alier et al. 2010; Haberl et al. 2011; Brand and Wissen 2013), with the renewal of sustainability thinking through the concepts of innovation, transition and transformation. Not critically analysing the modern economic world system and its environmental consequences, supporting the illusion that social–ecological adaptation can happen through further—ecological—modernisation of the industrial system (Mol 2001), or mainly through market-based governance and technological innovation, had as non-intended consequences delaying the necessary renewals of sustainability policies. Many processes required for environmental sustainability have not yet started; with the Millennium Ecosystem Assessment in 2005 it became clear that the environmental situation has hardly improved.

Indicators of the illusionary policies and practices in international agreements and global policy programmes are:

- the normative reasoning, with soft and abstract formulations of necessary value and behaviour changes, appealing to the great community of humanity, to our common interests, or the fate of humankind, in the rhetoric of policy and diplomacy;
- or the reasoning in terms of win–win situations, without addressing necessary transformation conflicts;
- or insufficient system analyses of economic and ecological systems;
- or the neglect of structural power relations in the analyses and of vested interests of powerful global actors.

Asymmetrical relations of power and domination, relations of production and social and economic inequalities in the globalised economic system, accumulation regimes and social–ecological regimes of natural resource use: these remain weakly analysed themes in governmental policies and government-dependent policy analyses. They cannot be corrected with new data and empirical knowledge, although empirical knowledge about the ineffectiveness of environmental policies is important for further improvements. Shortcut combinations of normative ideas or visions and empirical information in governance processes can only be corrected through reflexive knowledge processes: critically reflecting the normative premises and the neglected aspects of knowledge and governance practices, introducing in the discourse of environmental governance knowledge from critical theoretical analyses and reflections of the political and economic systems, of global social and environmental change, of structurally and systemically reproduced social and economic inequalities in the global economy as economically and ecologically unequal exchange (Rice 2007). Experiences with, misconceived forms of transformative collective action, or of cooperation of governmental and non-governmental organisations in the policy processes need to be assessed and reflected.

Pre-analytic visions, worldviews and paradigms as normative knowledge components directing scientific research are meanwhile widely discussed and reflected, but not further knowledge filtering processes that were less evident and more difficult to understand: the reflection and assessment of different knowledge forms and sources, the institutional fine-tuning of knowledge production through disciplinary specialisation of research, and the separation of the different knowledge processes of production, dissemination, transfer and application of scientific knowledge. For the environmental governance processes the insufficient review and reflection of knowledge practices results in obfuscation of alternative solutions and blocking of collective learning processes.

Ways to come out of the power and knowledge related dilemmas of inconsequent views and practices in environmental science and policy include: changes of the knowledge transfer processes, broadening of the knowledge bases for environmental agency through interdisciplinary knowledge integration, new forms of validation and sharing knowledge, new forms of review and evaluation of knowledge and governance practices, temporal structuring and phasing of the processes, and more general: developing of new knowledge cultures in the environmental discourse, in research and governance. Various principles for knowledge exchange and negotiation have already been formulated, though not yet widely practised: societal agenda setting, interdisciplinary research, collective problem framing, knowledge integration in research and policy processes, the changing research practices in sustainability science referring to complex systems and processes, working with a plurality of perspectives, extended peer review, effective dialogue processes, stakeholder participation, new norms for dealing with dissent and controversy, dealing with uncertainty and diversity of values (for example, the discussion of post-normal science: Funtowicz and Ravetz 1993), more transparent metrics for evaluation (Cornell et al. 2013). These discussions show potentials for successive improvement of governance. They require further systematisation with regard to principles and practices of interdisciplinary knowledge sharing and integration, which seem more developing through trial-and-error procedures than through epistemological reflection. With such reasoning began the search for other knowledge practices that can compensate the inherent weaknesses of specialisation-based expertise and knowledge use in the knowledge bridging between science and governance.

Through broader knowledge bases and more inclusive knowledge practices, it becomes possible to identify and deal more systematically with the relations of dependence, interdependence and interaction in governance processes: with interactions between social and ecological systems, between causes and consequences of action and decision-making, between governance systems and external or contextual factors and processes. This broadening of knowledge practices can help to open ways of renewing and improving global environmental governance, its effectiveness, efficiency and legitimation; but with that the governance processes become more difficult.

Analyses of the contextual factors and their influence on the environmental governance process are rarely attempted in the specialised research and knowledge use that prevails in science and policy. *With the interdisciplinary analyses and the broadening of the perspectives governance problems appear primarily as knowledge problems: less as lack of knowledge and lack of power, but as making more systematic and adequate use of the knowledge available and developing strategies for global agency that reflect the power and dependence relations in politics, in the economy, and in science.* With this conclusion from the prior chapters begins the following discussion of possibilities of rethinking and renewing global environmental governance and building global agency.

10.2 Advancing from Diagnoses of Deficits to Renewals of Global Environmental Governance

A *theoretically differentiated terminology* for processes related to global environmental governance in the perspective of socio-ecological transformation includes the terms of management, policy, politics, governance, and the connected concepts of cooperation, integration, institution and regime building, regulation, social and environmental change, adaptation, mitigation, transition and transformation, nature–society interaction, human and societal relations with nature, colonisation of nature. With this differentiation of theoretical concepts and processes analysed with these unfold more systematic analyses of complex interacting systems, and their transformation supported through environmental governance. The *logic of knowledge use in socio-ecological transformation* unfolds with these concepts, most of them abstract and requiring specification to be applied in governance processes. The concepts introduced in the prior chapters, in analyses of global environmental governance from different perspectives, show possibilities of integrating knowledge across the boundaries of natural and social sciences, for the critical analysis of environmental governance (Barry and Born 2013); such knowledge integration is still rare in environmental research, developing mainly in the interdisciplinary fields of human, social and political ecology. Two main difficulties in the renewal of global environmental governance include the dealing with incompatible views of governance and the practices of inter- and transdisciplinary knowledge integration.

1. *Differing views about global governance, its limits and its renewal.*
   Reasons given for lacking success and limits of global policies and governance include: complexity of ecological processes and problems, short-term thinking and action in politics, institutional fragmentation, time pressure, incompatible interests of actors and institutions or states, lack of leadership, vested interests of powerful players, not negotiable issues, unidirectional knowledge transfer instead of knowledge sharing between science and governance, difficulties to deal with expert controversies in the management of knowledge for public policies. These reasons include important, but not all important arguments for improvements; selectivity of knowledge practices and inefficiency of policy and governance processes are not critically analysed, the complexity and difficulty of problems and their solutions simplified, reifying them in epistemic descriptions as that of "wicked problems and clumsy solutions (Brooks and Grint 2010). To move beyond such reasoning, and beyond "proverbs of administration"

(Simon 1946) in the sense that one can always find generalised principles that explain success or failure, seems necessary in the further discussion. This can happen through changing practices of knowledge creation and use for governance, developing broader and different forms of knowledge use, reducing the weight of normative reasoning, beliefs and worldviews. Simon himself made first steps in the direction of interdisciplinary knowledge synthesis for administrative science, but his and most policy research after him did not go far enough to uncover systematically causes for the slowdown, limited success or failure of policies and governance regimes.

The guiding idea for rethinking environmental governance and dealing more adequately with its complexity, emerging in the preceding chapters is: *the development of global environmental governance is part of the overarching and longer processes of sustainability transformation.* In this perspective, the governance processes can be re-structured:

- from normal and routinised views of policy processes, the limitation of public policy and its institutionalisation
- to broader governance processes that include connections across various scale levels, and connections between political, economic and ecological processes.

This is not a new idea; the connection between different processes was articulated in the governance debate many times. Yet, the differing interpretations, the inconsistent use of the notion of sustainability, and simple views of limits of global policy processes devalue the knowledge work for sustainability; it is not based on theoretical systems analyses of the interacting social and ecological systems, and in the governance practice not supported through processes of consensus building, institutional reforms, and trans-political strategies of dealing with global change. The double rethinking global environmental governance—theoretically as connecting to more encompassing and other than policy processes, and practically for its renewing—are elaborated further in the discourse of social–ecological transformation, in terms of conceptual revisions, searching adequate policy framing concepts, building of transformative action capacity and transformative action groups, and connections between governance and other social and ecological processes.

The theoretical reconstruction of interaction processes in social and ecological systems affecting environmental governance is guided by the concept of social–ecological transformation for which the connecting and overarching processes have been formulated in Chapter 7 in the framework of nature–society interaction. This broadening directs the analysis and discussion to the core component, the theoretical system analysis of coupled social–ecological systems. Not yet discussed are further components of revising and reframing policy and governance processes, especially the broad debate about inter- and trans-nationalisation of politics, states, social and power structures that affect environmental governance (for example, Aretxaga 2003; Brand and Wissen 2013; Taylor 2017).

In this broader perspective the changes of environmental governance can be described theoretically in their dependence from changing societal structures and relations with nature with which practices of environmental action and governance need to be matched. The matching requires to find out, which social and ecological processes can be influenced, changed through governance and how, which not. Thus, environmental governance appears as a socially and ecologically embedded process which needs to become reflexive to become effective, the reflexivity implying such knowledge processes as theoretical reflection, interdisciplinary knowledge synthesis, collective learning and knowledge sharing. Structures to be taken into account in the governance and regulation processes include the structures of economic and political systems; the social practices connected with environmental governance include a variety of forms of individual and collective social action: individual changes of consumption and lifestyles; collective action in national, regional and international policy processes, in social movements and civil society development; cooperation between science and policy and new forms of knowledge production and application. Together the changes to be considered for and in environmental governance sum to a complex picture of long-term transformation of modern society and its relations with nature, of a transition to global sustainability.

Global environmental governance seen as a policy and power dominated process (Chapter 4, Table 4.4) does not include all processes and changes of which governance is a part or connected to. It shows in this conceptualisation a view from inside the policy processes, and only inexactly and selectively the limits of global governance and agency resulting from the contexts and the coexistence of contrasting policies. The research and reviews do not sufficiently account for the extra-political and extra-systemic contexts and influences upon environmental governance processes. The first critical question is about the limits of environmental governance, which includes the problems of knowledge creation and use.

The scientific research and publications on the limits of environmental governance cover many aspects: global environmental governance and earth system governance (discussed in Chapter 4), international governance, national and local governance, different forms of state, market and community based approaches, reforms of environmental governance (Oberthür and Gehring 2004; Chambers and Green 2005), institutional development (Paavola 2001), institutional forms and innovations (Esty and Ivanova 2002), collaboration and participation (Margerum and Robinson 2016), governance of complex systems and long-term governance (Underdal 2010), political economy of environmental governance (Newell 2008; Brand and Wissen 2013), state capacity and global environmental governance (Lenschow et al. 2016), conceptualisation and construction of global environmental governance (Ford and Kütting and Cerny 2015), and further specific aspects. These examples, not discussed further here, show the limits of specialised research on environmental governance requiring the development of other, broader, and theoretical knowledge syntheses.

2. *Inter- and transdisciplinary cooperation, knowledge exchange and integration*, the most important progress in the past decades, begin to influence governance processes—slowly, as could be seen in the discourse of Earth system governance. For the discussion here the limits and the deficits of governance need to be described in the perspective of social–ecological research, where strong interdisciplinary forms of integrating natural- and social-scientific

research emerge, and where the interaction between social and ecological systems is in focus. This is still not widespread interdisciplinary practice in environmental research; cooperation and knowledge integration across the boundaries of the natural and social sciences is badly developed, as discussed in the prior chapters, the publications referred to above indicating this again. The literature touches knowledge questions, but the focus is on policy, power, organisation, on institutional relations and processes. Critical and systematic interdisciplinary analyses of the limits and blockages of global environmental governance—combining system analyses of the modern capitalist world system and analyses of the interaction between society and nature, social and ecological systems—are not yet systematically used in the governance practices. Besides the reasons and forms of knowledge selection for environmental governance discussed above, the methodological and theoretical difficulties of such research and knowledge integration and its translation for the political knowledge practices impede the process. Also in interdisciplinary approaches as actor–network theory, and in critical environmental, political- and social–ecological research and reflection, differences and controversies can be found in the construction of global environmental governance: whether it requires—insofar as power relations are to be discussed—globally coordinated, hierarchical and top-down approaches, or participatory, networked and decentralised, multi-scale approaches. Or whether it requires approaches that start form subjective constructions of environmental governance in social, cultural or ethnic groups, resulting in actor-centred approaches, institutional diversity, networked and nested local resource use regimes.

Connecting perspectives and analyses of power relations and social emancipation, the variations and combinations of governance approaches become still more diverse and controversial. With the emancipatory approaches and the broadening of governance to transformative action come up questions discussed as the development of transformative science,—literacy and—agency; the subjects of social–ecological transformation such as transformative action groups and other social and political groups; the forms of changing power relations, and the forms of democratic decision-making and legitimation. Rethinking and renewing of global environmental governance in social–ecological and transformative perspectives cannot yet build on advanced knowledge integration or identify final solutions to the deficits and limits of governance; the progress is only a further step in the longer learning processes on the way towards sustainability, with ideas how to renew and change global environmental governance in the foreseeable near future. Moderate as this is in terms of knowledge production, integration and use, it implies dramatic forms of social, political and institutional change, beyond the scope of policy reforms. Regarding the literature discussed so far in this chapter and in Chapter 4, it can be asked: what is not considered in the environmental governance discourse, which knowledge is ignored in the policy and resource management processes? Attention should be directed to processes that are considered as not manageable and controllable in international environmental regimes. Processes that cannot be managed or influenced through policy and governance seem to be irrelevant for the environmental governance debate, although they may influence and limit the governance processes. In contrast to disciplinary specialisation of policy research where this question is ignored, the development and improvement of environmental governance and global agency depends on the analysis of non-manageable processes in social and ecological systems "beyond the horizons of governmental policy": how can these be connected with managed, mediated and controlled processes?

Interdisciplinary cooperation and knowledge integration in environmental research exist since long time, in projects and institutions as that described in (Box 10.1), dealing with research, knowledge synthesis and transfer for sustainable development and environmental governance. These institutions, nodes in the global networks of environmental governance, show the development of inter- and transdisciplinary ecological discourses in their activities—and how dispersed, multi-paradigmatic, controversial the development is. Natural-scientific research and knowledge are still prevailing, used in weak forms of interdisciplinary integration of social-scientific knowledge, from which more complex theories of society and nature remain excluded. This goes together with more invisible and implicit assumptions and premises such the unspoken reality taken for granted, that changes and adaptations to environmental change can be achieved without changing the societal and economic systems, in the modern economic world system and its political world order. As a consequence of that the term of transformation adopts various diffuse meanings, is deprived of its critical implications of system-transformation.

The term "transformation" has within the past decade advanced to a main term in discussing sustainability, however, as the term sustainability before, it is used in abstract and diffuse forms, not identifying the requirements of transformation of social and ecological systems. Transformation may refer to value orientations, behaviour practices, lifestyles and consumption, technologies, globalisation, temporal and spatial dimensions of action—only with its specification as social–ecological transformation modern capitalism, industrial society, the interaction and regulation of nature–society relations from local to global scale levels were included in transformation towards sustainability.

Box 10.1 Practices of interdisciplinary cooperation and knowledge integration in environmental governance—examples of global "think tanks"

Knowledge-generating, synthesising and transferring institutions for supporting sustainability policies and environmental governance:

the International Institute of Sustainable Development (IISD), Winnipeg et al.; the Potsdam Institute for Climate Research in Germany; the Stockholm Resilience Center and the Resilience Alliance; the Center for Global Environmental Research at the National Institute for Environmental Studies in Japan; the Socio-Environmental Synthesis Center (SESYNC) at the University of Maryland in the United States—describing on its homepage the pragmatic maxim paradigmatically as "bringing together the science of the natural world with the sciences of human behaviour and decision making to find solutions to complex environmental problems"

Interdisciplinary knowledge cultures of "down to earth"-pragmatism:

These institutions at the interface of science and governance practice different strategies of interdisciplinary environmental research and

> knowledge transfer; in their knowledge practices dominates natural-scientific research and knowledge; less supported and used is social-scientific knowledge (especially theoretical knowledge); environmental research and knowledge integration is seen as pragmatic, applied science for the purpose of decision-making, management, policy or governance; as a consequence of the traditions and knowledge practices in academic and specialised environmental research, knowledge production relies on empirical research and data, rather neglecting the relevance of theoretical knowledge and synthesis across the boundaries of natural and social sciences; there is no consensus about principles and methods of knowledge integration, synthesis, and application—such consensus could be useful for the practice of knowledge application in global environmental governance, but can only develop when there is broader consensus in the scientific community (as, for example, in climate governance)
>
> *Source* information from the homepages of the institutions

The selective knowledge practices in environmental governance reduce governance, transformation, sustainability to goals to be achieved in a short time, under given circumstances, through politics, power, knowledge and influence. Outside of the horizons of governance debates remain the changes of societal and economic systems, of social–ecological systems and social practices that are to be considered in long-term transformations to sustainability, with time horizons of several generations or hundreds of years. Such change is beyond the scope and time frame of policy and planning processes, shifted to discussion in future studies, where it is detached from the practices and necessities of action and decision-making. The limited temporal perspectives of policy and planning processes are that of policy and decision-making process, where the near future is discussed, consensus and support of actors and social groups sought.

Beyond the near future is the time horizon for transformations to sustainability, beyond the range of political planning. It appears as the future in terms of evolutionary changes of complex and coupled social and ecological systems, for which, as practical experience says, no effective action strategies can be developed. Yet, the long-term changes influence governance processes in many ways, as it became clear with climate change or with globalisation, where the long-term perspectives and consequences of social and environmental change determine the possibilities of environmental governance that can only be identified with a more detailed phasing and structuring of governance. In the long-term perspective of social–ecological transformation it becomes necessary to modify strategies in the progress towards sustainability. Modifications are caused by new knowledge, experiences and insights, that require developing of capacities and capabilities by the actors in terms of collective learning, transformative literacy and agency, dealing with structural components of societal systems that block transformation, and with second-order regulation or capacities of regulation of regulations that are components of the complex transformation process. The quality of governance strategies is not only the mobilisation of people and support, but to keep the future open for further change, not foreclosing necessary options and paths for achieving sustainability.

3. *Concluding from these considerations* and the arguments developed in the preceding chapters, the most important component of sustainability strategies is: to separate between processes that can be regulated and included in global governance and agency and their reach, and autonomous social and ecological structures and processes that cannot be directly regulated and influenced through governance decisions; these can only indirectly be influenced, navigated, modified. The processes relevant to global environmental governance can, theoretically seen, be described in three kinds: (a) governance processes, (b) connected complex processes that can be regulated, and (c) connected complex processes that cannot be regulated as whole processes, only in successive parts.

   a. *Processes that are part of governance regimes* and connected with global environmental governance as an institution-building process include the environmental and sustainability policies and respective global environmental regimes, the governance and regulation of the global economic system, the framing institutions and processes of decision-making about sustainability and the respective normative orders in terms of participation, citizen and human rights.

b. *Complex processes in social and ecological systems that can be managed,* regulated, and connected with global environmental governance include the social processes of economic production, accumulation regimes, global exchange and trade, distribution and sharing of natural and other resources, energy regimes, ownership rights, consumption and lifestyles, demographic processes, and the societal relations with nature. Many of these processes can be described as part of social reproduction processes.

c. *Complex processes that cannot be regulated as whole processes, only indirectly and in successive phases, through building of composite, phased, long-term strategies of socio-ecological transformation include:* the development and transformation of modern society as a whole, the development and change of global ecological systems and processes, including material cycles, climate change, biodiversity, and the long-term processes of biological evolution. For all of these processes the terms of management, governance and regulation are used too, but become, in the theoretical perspective followed here, inadequate. The complex systemic processes can be influenced, modified, manipulated by humans, but the terminology of change, governance and regulation needs to be differentiated and specified to describe the qualities of these complex processes, including differentiations in temporal and spatial scales.

Differentiation of time scales: With the long-term time horizon of social–ecological transformation to a sustainable society, discussed from different perspectives in this book, the view of global environmental governance and agency becomes another one, beyond the limited views and short-term perspectives of policy-oriented research and reform debates. These remain debates about the near future, the initiation of changes that can be foreseen and planned for, mostly shorter than a generation, but continue as transformation processes of several or many generations, inherent in sustainability thinking. The time frames do not only become longer, but differentiated in several qualitatively different, connected subsequent phases. Beyond that what can be planned for (for which the methodological approaches in future studies become relevant, presently mainly in use scenarios), the time regimes of sustainable development become more differentiated, including nonlinear, bifurcating and cyclic changes. With the transformation debate the perspectives of development and change become that in the historical dimensions of *longue durée*. Projecting this idea in the formulation of future processes of societal change and transformation, requires the use of methods and the results of long-term historical analyses of the trajectories of societal development, as they investigated for historical socio-metabolic regimes and for the modern society in world system analyses and world ecology. Possible future development paths need to be constructed to include the possibilities of unforeseeable changes in nature and society, ruptures and crises, nonlinear and bifurcating development.

Connected with the typology of governance-related processes described above in three forms according to the governability of societal changes, a *conceptual framework of different phases of societal development,* projected in future trajectories of achieving sustainability, is necessary to assess the gradual achieving of sustainability. The framework can be developed in terms of theoretical and methodological requirements for transformation analyses:

a. *phases of change in the near future*: these can be described with the methods and perspectives of public planning and resource management;
b. *phases of change in the long-term future*: these can be described on the basis of multi-scale and global scenario-analyses, and the methodology of future studies;
c. *phases of change in the distant future that are beyond the time scales of scena*rios: these processes of the very long transformation (in centuries of duration) can only be described in theoretically formulated perspectives that make use of knowledge and insights about former great transformations in human history. At this level of thinking the future the theoretical knowledge of social ecology can be used in discussions of the renewal of global environmental governance.

The differentiation of transformation processes in long-term perspectives is required to assess the advances achieved that are in preliminary forms described in scenario-based analyses in views of the distant future as open but influenced by decisions in earlier phases. In a systematic differentiation of these three temporal perspectives, knowledge gaps and limits of environmental and governance research can be more concretely discussed in the renewing of global environmental governance.

10.3 Framing and Contextualisation of Global Environmental Governance—Integrating New Knowledge

Global environmental governance and the integration of national and global environmental policies began in the 1980s with climate policy and the discourse of sustainable development. A symbolic date is the UNCED conference in Rio in 1992, resulting in the global action programme "Agenda 21". Since then global environmental governance developed with more and more actors, organisations and institutions and new international environmental regimes. The broadening of the spatial, temporal and knowledge horizons, the networked and integrated strategies of action brought increasing complexity and density of environmental regulation with—so far—modest results in regulating global environmental change, due to reasons and causes discussed in this book.

In the course of three decades, the global governance processes had difficulties of finding support; repeated relapses happened in the important fields of climate change, biodiversity maintenance and sustainable development. This can be seen as reflecting the complexity of global policies and relations, of the regulation of complex adaptive systems in society and nature. It shows, however, also the controversial nature and forms of global governance in science and in politics, articulated in continuing controversies about the construction and organisation of the governance processes. Governance deficits in terms of effectiveness and efficiency can be seen in the vague goals and principles, for example in the "green economy" strategy of the UNEP; in the lack of effective strategies of social transformation, and in the exclusion of important social components from the governance processes, such as conflict regulation. For the conflicts between organisations, institutions, powerful groups, countries and governments that come up in the governance processes mitigation strategies are badly developed, as the integration of social, economic and ecological transformation processes to sustainability. With the process of economic globalisation many processes of change affecting environmental governance came under the influence of the economy and the markets, were reduced to technological change and innovations to be realised with another accumulation regime or another industrial revolution. In the economic development strategies much of that what is required for sustainable development ended in the traps and dead ends of illusionary policy reforms that prevent or shift the reduction of natural resource use, redistribution and sharing of resources.

The consequences of the crisis of environmental regulation and governance were described in Chapter 4 in more detail as a knowledge-related crisis. The significance of scientific knowledge for environmental governance has increased and was simultaneously challenged, at a time when nearly all other forms of knowledge relevant to human history have been dominated through science. The revival of non-scientific, local and practical ecological knowledge and the discourse of transdisciplinarity indicate a late crisis in scientific knowledge production. The intra-scientific changes in the organisation of research and knowledge transfer, reflected especially in the discourse on new knowledge production and transdisciplinarity, showed more clearly the limits of knowledge, the fragmented and incomplete knowledge generation through specialised research, with competing and contradicting research results and interpretations dating back to disciplinary, methodological, theoretical and epistemological differences, to different worldviews, paradigms and normative assumptions influencing research. Science became, using a Weberian term, "dis-enchanted". Two changes happened so far in environmental research as reaction to the knowledge crisis: (a) interdisciplinary cooperation and knowledge synthesis, which can be seen as attempts to deal with the social limits of knowledge, the "specialisation syndrome", and (b) participatory research and decision-making in the governance process, that can be seen as seeking consensus and developing forms of societal agenda setting. Both changes did not generate immediate success in terms of improving environmental governance, but they made visible the difficulties of organising global governance and creating global agency.

Among the difficulties and deficits of global environmental governance described at the beginning of the chapter, many indicate unsolved problems of knowledge use, of production, dissemination and application of knowledge:

- the *complexity of environmental problems, social and ecological systems and their interaction* require hitherto unknown forms of interdisciplinary cooperation, in exemplary ways visible in climate research and policy;
- the *processes of scientific knowledge production*, the epistemic problems of insufficient, uncertain and controversial knowledge, of verification of knowledge, the division of labour and specialisation, and the growth of specialised knowledge result in increasing difficulties of knowledge integration;
- the *social complexity of knowledge practices in the chain of knowledge production, transfer and application and political action*, the multitude of actors involved, the multi-scale action, the manifold processes of interest-matching, consensus building, institution-building, legitimation and coordination of global environmental governance make science communication, negotiation and action complicated;
- the epistemologically and methodologically insufficiently structured forms of knowledge bridging, integration and synthesis in environmental research and governance add up to the difficulties.

Part of the knowledge-related problems and difficulties is the interaction and blending of scientific, political, social and cultural functions and processes in the creation of global agency: in each phase of the process are involved scientific and knowledge problems, political and power problems; social and cultural problems with different worldviews, values, interests; problems of communication, of building environmental literacy and capacity, institutions and international networks.

Concluding from these observations improving of global environmental governance cannot be done through monitoring, evaluation and policy research alone, where only a limited number of factors and variables are accounted for and knowledge is selected with regard to the functioning of single political institutions and processes, ignoring the broader context of environmental governance. When the social, ecological and knowledge contexts of governance are in view, the problems and difficulties turn out to be such of finding adequate forms of dealing with the complexity and of reducing this complexity. The discussion of, the developing and improving global environmental governance differentiates into a variety of perspectives and questions (summarised in Table 10.1).

Regarding knowledge production and application, the ways to overcome the limits of global environmental governance described in terms of insufficiency, inadequacy, incoherence and selective knowledge use can be described in different forms: to make better use of interdisciplinary and social–scientific knowledge available in theoretically codified forms. This includes knowledge integration and synthesis, the

Table 10.1 Problems of developing environmental governance

1. *The knowledge basis of environmental governance:* scientific and other forms of knowledge (local and practical knowledge; normative knowledge)
2. *The contextual processes and external system-dependence of governance:* the social organisation of society and its functional differentiation (political and economic systems, science, culture); the functioning of ecosystems; the coupling of social and ecological systems
3. *The institutional architecture of governance:* the continuous search for improved organisation; hierarchical or networked governance; parsimony or redundancy; polycentric; parallel structures; the shifting of organisational mechanisms form goals and structures to processes and temporal phasing
4. *The processes of development and change of governance:* path-dependence and path-transformation, trends and mega-trends, nonlinear development
5. *Decision-making and collective action:* which forms of decision-making and action does the knowledge complexity require (e.g. double-loop learning processes; anticipation, prognoses; second-order regulation; dealing with unexpected events and changes; criteria and forms of consensus building, conflict mitigation, legitimation, democratic decision-making in the global arena)

Source own compilation

organisation of governance-related research, and the knowledge to be used in environmental governance.

1. *Knowledge integration and synthesis includes different processes of*

- integrating results from ecological research and analyses of ecological systems in governance,
- integrating results from social–scientific environmental research and analyses of social systems in governance,
- using theoretical knowledge from theories of modern society and interaction of society and nature,
- creating methodologies for interdisciplinary knowledge generation, synthesis, transfer and application.

2. *Governance-related research is confronted with different difficulties of knowledge transfer and sharing:*

- The environmental knowledge crisis is not primarily caused through lack of knowledge about the functioning and the processes in complex systems, but through controversies and selective use of knowledge required for dealing with the governance crisis.
- The social–ecological analysis of the interaction of modern society and nature shows, that the regulation of the interface of nature and society includes more processes beyond environmental policy.
- The social–ecological analysis of nature–society interaction (discussed in Chapter 8) shows global social–ecological transformation that includes processes in the social and ecological system that can only partially or indirectly influence through environmental governance and environmental regimes.

3. *The knowledge to be integrated into governance processes includes different fields of specialised research:*

- integrating results from ecological research—the functioning of ecosystems,
- integrating results from social-scientific research—resource use in social systems,

- theoretical knowledge about the interaction of nature and society—the planetary boundaries of resource use,
- epistemological reflection of interdisciplinary knowledge production, synthesis, transfer and application.

The main knowledge problem is the integration of results from natural- and social-scientific research; in interdisciplinary research it is rarely analysed in terms of epistemological or ontological problems of knowledge integration (as suggested, for example, by Barry and Born 2013: 191ff), more often in social forms of collaboration, organisation and communication, and science communication.

Regarding these problems of knowledge integration and application global environmental governance becomes more complex and complicated than hitherto; furthermore, it is confronted with long processes that stretch in the distant future. Environmental governance is no longer a process in conventional forms of decision-making, planning, formulation and implementation of policy programmes and environmental regimes; it becomes a process subjected to a temporal rationality of phasing, temporal structuring and continuous restructuring of sequential processes. The methods of future studies and analyses of future perspectives of transforming the capitalist world system (Brie 2014; see Chapters 8 and 9) become important for creating knowledge when no data and experiences are available. The flow and the passing of time, the continuous transformation of future into the present and past time, direct and determine the dynamics of environmental governance where the knowledge used changes continually.

10.4 General Conclusion: The Present Situation—Potential Improvements and Unsolved Problem in Social–Ecological Transformation

The analyses and theoretical reflections in this book indicate the following conditions of a renewal of global environmental governance:

1. *Suggestions for improving environmental governance through institutionalisation of new knowledge practices* (see Cornell et al. 2013): societal agenda setting, collective problem framing, integrative research processes, working with a plurality of perspectives and parallel strategies of knowledge use, extended peer review, effective dialogue processes, stakeholder participation, new norms for dealing with dissent and controversy, dealing with uncertainty and diversity of values, more transparent metrics for evaluation; inserting social–ecological knowledge in governance processes; organisation of global environmental assessments with participation of stakeholders; multi-stakeholder councils.

2. *Suggestions for reorganising organisational processes of environmental governance:* new transformative subjects (pluralistic "transformation action groups"); innovation networks; coalitions of governmental and non-governmental organisations; public–private partnerships; private governance; intergovernmental governance; coordinated multi-scale governance; improving intergovernmental governance and cooperation of states in the international arena.

3. *Suggestions for contextual strategies to support socio-ecological transformation:* differentiation of development strategies at national and regional levels, including the modification industrialisation as the dominant development strategy through combination with other forms of economic development and natural resource use (partial industrialisation, development of rural economies, transitional economies); developing transformative science and the building of transformative capacity.

4. *Critique of ideas to bypass socio-ecological transformation:* technical fixes and technology transfer; market-based governance and globalisation; solar geo-engineering and genetic engineering.

Global environmental governance is continually confronted with the problems of building a new political world order, one with more equal power-relations in and between the global north and south, the international institutions, governments, and non-governmental organisations. In environmental policy analyses, where the relations and the interaction between national and international policies are analysed, the complexity and the difficulties of building global agency have become visible in manifold conflicts (political, economic, cultural), in asymmetrical power relations, in coordinating and integrating multi-scale governance across local, national and global scales. Global environmental governance cannot be improved without building a new world order and development of institutions to strengthen global agency that is necessary for the complex processes of social–ecological transformation.

References

Agrawal, A. (2003). Sustainable Governance of Common Pool Resources: Context, Methods, and Politics. *Annual Review of Anthropology, 32,* 243–262.

Ascher, W., Steelman, T., & Healy, R. (2010). *Knowledge and Environmental Policy: Re-imagining the Boundaries of Science and Politics.* Cambridge, MA: MIT Press.

Atkinson, R., Terizakis, G., & Zimmermann, K. (Eds.). (2011). *Sustainability in European Environmental Policy: Challenges of Governance and Knowledge.* New York: Routledge.

Aretxaga, B. (2003). Maddening States. *Annual Review of Anthropology, 32,* 393–410.

Barry, A., & Born, G. (Eds.). (2013). *Interdisciplinarity: Reconfiguration of the Social and Natural Sciences.* London and New York: Routledge.

Becker, C. D., & Ostrom, E. (1995). Human Ecology and Resource Sustainability: The Importance of Institutional Diversity. *Annual Review of Ecology and Systematics, 26,* 113–133.

Bizer, C., & Cyganiak, R. (2009). Quality-Driven Information Filtering Using the WIQA Policy Framework. *Web Semantics: Science, Services and Agents on the World Wide Web, 7*(1), 1–10.

Bown, N., Gray, T., & Stead, S. (2013). *Contested Forms of Governance in Marine Protected Areas: A Study of Co-management and Adaptive Co-management.* London and New York: Routledge.

Boyd, E., & Folke, C. (2011). *Adapting Institutions: Governance, Complexity and Social-Ecological Resilience.* Cambridge, UK: Cambridge University Press.

Brand, U., & Wissen, M. (2013). Crisis and Continuity of Capitalist Society-Nature Relationships: The Imperial Mode of Living and the Limits to

Environmental Governance. *Review of International Political Economy,* *20*(4), 687–701.

Brie, M. (Ed.). (2014). *Futuring: Perspektiven der Transformation im Kapitalismus und darüber hinaus.* Münster: Westfälisches Dampfboot.

Brooks, S., & Grint, K. (Eds.). (2010). *The New Public Leadership Challenge.* Houndmills, Basingstoke, UK: Palgrave Macmillan.

Chambers, W. B., & Green, J. F. (2005). *Reforming International Environmental Governance: From Institutional Limits to Innovative Reforms.* Tokyo: United Nations University Press.

Cornell, S., et al. (2013). Opening up Knowledge Systems for Better Responses to Global Environmental Change. *Environmental Science & Policy, 28,* 60–70.

Daviter, F. (2015). The Political Use of Knowledge in the Policy Process. *Political Science, 48,* 491–505.

Dimitrov, R. (2006). *Science and International Policy: Regimes and Nonregimes in Global Governance.* Lanham: Rowman and Littlefield.

Esty, D. C., & Ivanova, M. H., eds. (2002). *Global Environmental Governance: Options & Opportunities.* New Haven, CT: Yale Center for Environmental Law and Policy.

Fischer, F. (2000). *Citizens, Experts, and the Environment: The Politics of Local Knowledge.* Durham and London: Duke University Press.

Funtowicz, S., & Ravetz, J. (1993). Science for the Post-normal Age. *Futures, 25*(7), 739–755.

Garard, J., & Kowarsch, M. (2017). Objectievs for Stakeholder Engagement in Global Environmental Assessments. *Sustainability, 9,* 1571. https://doi.org/10.3390/su9091571.

Haberl, H., et al. (2011). A Socio-metabolic Transition Towards Sustainability? Challenges for Another Great Transformation. *Sustainable Development, 19,* 1–14.

IAASTD. (2009). Agriculture at a Crossroads. *International Assessment of Agricultural Knowledge, Science and Technology for Development* (IAASTD). Washington, DC: Island Press.

Jones, H., Jones, N., Shaxson, L., & Walker, D. (2012). *Knowledge, Policy and Power in International Development: A Practical Guide.* Bristol: The Policy Press.

Kovacic, Z. (2017). Investigating Science for Governance Through the Lenses of Complexity. *Futures.* https://doi.org/10.1016/j.futures.2017.01.007.

Kütting, G., & Cerny, P. G. (2015). Rethinking Global Environmental Policy: From Global Governance to Transnational Pluralism. *Public Administration, 93*(4), 907–921.

Lemos, M. C., & Agrawal, A. (2006). Environmental Governance. *Annual Review of Environment and Resources, 31,* 297–325.

Lenschow, A., Newig, J., & Challies, E. (2016). Globalization's Limits to the Environmental State? Integrating Telecoupling into Global Environmental Governance. *Environmental Politics, 25*(11), 136–159.

Margerum, R. D., & Robinson, C. J. (Eds.). (2016). *The Challenges of Collaboration in Environmental Governance: Barriers and Responses.* Cheltenham, UK and Northampton, MA, USA: Edward Elgar.

Martinez-Alier, J., Pascual, U., Vivien, F. D., & Zaccai, E. (2010). Sustainable De-growth: Mapping the Contexts, Criticisms and Future Prospects of an Emergent Paradigm. *Ecological Economics, 69*(9), 1741–1747.

Millner, A., & Ollivier, H. (2016). Beliefs, Politics, and Environmental Policy. *Review of Environmental Economics and Policy, 10*(2), 226–244.

Mol, A. (2001). *Globalization and Environmental Reform. The Ecological Modernization of the Global Economy.* Cambridge, MA and London: MIT Press.

Newell, P. (2008). The Political Economy of Global Environmental Governance. *Review of International Studies, 34*(3), 507–529.

———. (2011). The Elephant in the Room: Capitalism and Global Environmental Change. *Global Environmental Change, 21,* 4–6.

Oberthür, S., & Gehring, T. (2004). Reforming International Environmental Governance: An Institutionalist Critique of the Proposal for a World Environment Organisation. *International Environmental Agreements: Politics, Law and Economics, 3,* 359–381.

Paavola, J. (2001). Institutions and Environmental Governance: A Reconceptualization. *Ecological Economics, 63,* 93–103.

Peterson, M. J. (2018). *Contesting Knowledge in International Environmental Governance.* London and New York: Routledge.

Prewitt, K., Schwandt, T. A., & Straf, M. L. (Eds.). (2012). *Using Science as Evidence in Public Policy.* Washington, DC: The National Academies Press.

Rice, J. (2007). Ecological Unequal Exchange: Consumption, Equity, and Unsustainable Structural Relations Within the Global Economy. *International Journal of Comparative Sociology, 48*(1), 43–72.

Simon, H. A. (1946). The Proverbs of Administration. *Public Administration Review, VI, 1,* 53–67.

Sörlin, S. (2013). Reconfiguring Environmental Expertise. *Environmental Science & Policy, 28,* 14–24.

Taylor, I. (2017). Transnationalizing Capitalist Hegemony: A Poulantzian Reading. *Alternatives: Global, Local, Political, 42*(1), 26–40.

Underdal, A. (2010). Complexity and Challenges of Environmental Governance. *Global Environmental Change.* https://doi.org/10.1016/j.gloenvcha.2010.02.005.

Young, O. R. (2011). Effectiveness of International Environmental Regimes: Existing Knowledge, Cutting-Edge Themes, and Research Strategies. *PNAS, 108*(50), 19853–19860.

Index

© The Editor(s) (if applicable) and The Author(s) 2019
K. Bruckmeier, *Global Environmental Governance*,
https://doi.org/10.1007/978-3-319-98110-9

287

The manufacturer's authorised representative in the EU is Springer
Nature Customer Service Centre GmbH, Europaplatz 3, 69115 Heidelberg,
Germany. If you have any concerns regarding our products, please
contact ProductSafety@springernature.com

Printed and bound by CPI Group (UK) Ltd, Croydon, CR0 4YY
23/04/2026
02095636-0001